THE
SCIENCE
OF
MILITARY POSTS,
FOR THE USE OF
REGIMENTAL OFFICERS,
WHO FREQUENTLY COMMAND
DETACHED PARTIES,
IN WHICH IS SHEWN

The manner of ATTACKING and DEFENDING
POSTS.

M. LA COINTE,
of the ROYAL ACADEMY at NISMES
1761

The Naval & Military Press Ltd

published in association with

FIREPOWER
The Royal Artillery Museum
Woolwich

Published by
The Naval & Military Press Ltd
Unit 10 Ridgewood Industrial Park,
Uckfield, East Sussex,
TN22 5QE England
Tel: +44 (0) 1825 749494
Fax: +44 (0) 1825 765701
www.naval-military-press.com

in association with

FIREPOWER
The Royal Artillery Museum, Woolwich
www.firepower.org.uk

In reprinting in facsimile from the original, any imperfections are inevitably reproduced and the quality may fall short of modern type and cartographic standards.

[iii]

THE FRENCH AUTHOR dedicates his work to the prince of Conti, having, as he says, acquired his knowledge under his highneſs in Piemont and Flanders.

❈❈❈❈❈❈❈❈❈❈❈❈❈❈❈❈

The following pieces, printed before the original, we think proper to give with the tranſlation, that the reader may ſee what value was ſet upon it in France.

A letter from marſhal count de Lautrec to M. la Cointe, captain of cavalry.

SIR,   Paris, Sept. 25, 1758.

I Return you your little treatiſe in manuſcript on the defence of military poſts. I read it with attention, and I preſume, if you make it public, it will be favourably received; for what you relate on this ſubject is very inſtructive, and young officers may from thence learn principles proper to direct them how beſt to preſerve poſts that are entruſted to them; which, by their ſituation

tion, are often of infinite confequence, and contribute much to the fafety of the camp, as well as to the army on its march; befides feveral other occafions wherein they may be of great utility.

This, Sir, is what I think of the little work on which you have confulted me; your application, and the zeal you fhew for the king's fervice, are equally commendable. I wifh that the court, knowing your merit and your talents, may not let them lie idle. You need not doubt, that for my own part, I fhall take all opportunities to recommend them; affuring yourfelf, Sir, that no body can be more difpofed than I am to oblige you on all occafions.

Le Marechal de LAUTREC.

The extract from the regifter of the royal academy at Nifmes,

Contains no more than an acknowledgment of M. la Cointe's being of their academy, and allowing him, after fubmitting his work to be examined by a committee of their *members*, to publifh it with his character of Academift in the title page.

Appro-

Approbation of Messrs. the committee of the academy of Nismes.

WE the undersigned, commissaries named by the royal academy at Nismes, to examine a work of M. la Cointe, intitled, *the science of military posts*, &c. certify, that we have read this work with attention. It has appeared to us the more useful and instructive for young officers, because no one has hitherto methodized that branch of the art of war, which is the subject of this book, into principles that may direct their practice; which is a matter of great importance, from the connection between it and the grandest operations of armies.

These motives engage us, in the name of the royal academy of Nismes, which has impowered us so to do, to permit M. la Cointe to take in his work the title of Academist of Nismes. In witness whereof we have given the present cer- certificate. Paris, Feb. 20, 1759.

Le marquis D'Aubais Menard.

Approbation of M. Belidor, brigadier of the king's armies, censor royal for the artillery and engineers of the academy of Berlin.

I Have read, by order of my lord the chancellor, a manuscript, intitled, *the science of military posts*, by M. la Cointe, formerly lieutenant of infantry, since captain of cavalry. This work, wrote with a great deal of care, method and erudition, comprehends the best maxims that can be given on the manner of fortifying and defending the advanced posts of an army; it will therefore be of great utility to young officers; as I know of nothing more instructing, or more proper to excite emulation. Given at Paris this 19th of Feb. 1759.

BELIDOR.

DEDICATION.

To the subaltern officers in the British army.

Gentlemen,

A Severe illness, in consequence of some hard campaigns, confined me for several months to my chamber; as much of this time as my disorder would permit, I began to employ in studying such authors as might be of use to me in my profession; among which, the little piece which I now present you with, seemed so well calculated to be the pocket companion of a young soldier, that I thought, though I was unable to do my duty in the field, I might in the mean time do an acceptable service to my brother officers, by recommending Mr. le Cointe's lessons to them in our own language.

I am, Gentlemen,

   Your sincere well wisher,

     and most humble servant,

       The Translator.

Translator's PREFACE.

THE translator had rather be blamed for a bad stile, or for copying the French idioms, than run the risque of changing the sense of the original by too much polishing: a sentence brought from one language to another, when relative to science, should be changed as little as possible, for fear of inverting the sense; however, in some places, especially in the geometrical part, he was obliged to help his author; but for the rest, as his intention was to give a faithful translation, whatever thoughts of his own occurred, he has given them by way of notes at the bottom of the page

As for the merit of the work itself, since the foregoing approbations certify it to be an original work in France, we apprehend it will be to the full as new and as useful to young military men in Great Britain, where the theory of the art of war ought to be the more carefully cultivated, as the happiness of our situation, and mode of government, give us, in comparison with our ambitious neighbours, but few opportunities of practising it.

CONTENTS.

Approbation of the work, by marshal Lautrec — Page iii
——————— by the academy of Nismes — — — v
——————— by M. Belidor vi
Translator's dedication
——————— preface
Introduction Page 1

CHAP. I.

Geometry necessary for an officer 13
Of the point and of the line — 14
Of angles — — — — — 19
Of triangles — — — — 21
Of surfaces — — — — 23
Of solids — — — — — 26
Of practical Geometry, i. e. to carry the foregoing rules into execution on the ground — — — — — 29

CHAP.

The CONTENTS.

CHAP. II.
Of the different works, with which posts may be fortified —— p. 36

CHAP. III.
Of the different ways of increasing the strength of posts — — — 49

CHAP. IV.
Of the necessary preparations to go on detachment — — — — 68

CHAP. V.
Of the march of detachments to posts 73

CHAP. VI.
Of the establishment of a body in a post 92

CHAP. VII.
Of precautions to be taken in a post, to avoid a surprize — — — 102

CHAP. VIII.
Of dispositions necessary to maintain a party in a post — — — 127

CHAP.

The CONTENTS.

CHAP IX
The defence of posts — — p 134

CHAP X.
Of the attack of posts — — 154
How to reconnoitre a post — — 155
Of the choice of soldiers — — 158
Of dispositions — — — ibid
Of guides — — — — 160
Of the March — — — 162
The attack of a common redoubt 163
The attack of intrenchments with a revetement — — — — 165
Of the passage of a ditch full of water
 168
Ways to counter-act the other contrivances — — — 170
The attack of a chateau, *or of a house*
 171
The attack of a village — — 174
Of seizing posts by stratagem — 187
Extract *from* M de la Croix *on the* petty war — — — — 206
Preparations for a march — — 207

The

The CONTENTS.

The use of infantry, and the utility of horse in a retreat — — p. 211

Care and precautions to be taken in towns, villages, and places of refreshment 213

Other precautions and measures for night-marches; attention to the fire-arms; and the essential custom for retreats 214

An useful maxim for rencounters, nocturnal and unforeseen attacks — 216

A stratagem commonly made use of by M. de la Croix — — 218

The advantage of night attacks, and the precautions to be taken in quarters 219

Some hints *and* observations *borrowed from* M. Saxe

Of war among mountains — 225
Of a country inclosed by hedges and ditches 227
Of the passage of rivers — — 230

INTRODUCTION.

THE ambition which animates our young military people is commendable, and becomes every day a gain to the ſtate.

Excited by examples, which they have before their eyes, to make their way to honours and favour, and convinced, that capacity and talents entitle them to ſucceed, they are leſs occupied with amuſements that waſte their time, and have more application

This idea has produced a happy change, and we now ſee more emulation and zeal than ever there was before; almoſt all officers ſtudy, almoſt all officers draw; and excepting a ſmall number, who look upon the ſervice as

a life

a life of independence, wherein they may take the liberty to neglect all the sciences; there are few who do not feel how advantageous it is to put themselves in the way of being known. The progress of reason has had an influence on all the arts; and to appear on the parade, and be master of the manual exercise, is no longer looked upon as the only merit necessary in a regimental officer, because we see these duties performed by soldiers ever so little disciplined.

A man that would advance himself, must study every branch that belongs to the art of war.

As the end that every one proposes, when they enter into a profession, is to advance themselves therein; the whole care of a young officer should be to instruct himself in that of war.

As I was designed for that profession from my earliest youth, I learned betimes, and have studied ever since, that part of mathematics which is most important for a young officer to know.

This

[3]

This application, and twelve years experience in the foot service, having caused me to make reflections on the fortifications of posts, to which regimental officers may be detached, I have given them to the world, persuaded, that they may be useful and advantageous to the King's service.

Another reason determined me thereto: having seen, in the different detachments that I made during the last war in Piedmont and Italy, how much a young officer, who has no idea of fortifications, is embarrassed, when he is ordered to intrench himself; I thought a book, containing principles by the help of which such works may be easily performed, and which would give, at the same time, the methods of defending and attacking such posts, would be a very great help to them.

No author, that I know of, has hitherto laid down these principles, so as to render them immediately useful to young officers; the most of them have seemed as if they intended only to give lessons to Generals, by writing profound treatises

treatises on the grand operations of an army, and as if they difdained to expatiate on fuch as they imagined were lefs important.

Chevalier Follard, and Chevalier de Clairac, are the only writers on the attack and defence of pofts of this nature; but the former, whom we may confider as the reftorer of the true principle of war, has touched them but lightly; and the rules, given by the latter, are fo connected with the greater works made in the field, fuch as intrenching of armies, lines of communication, and trenches, that they can be of little fervice to private or regimental officers.

The authors who have written after them, have gone no deeper into this part of the fcience, becaufe they did no more than either to copy or abridge the others, without ever entering into thefe particulars that the fubject is capable of. However, the fcience of pofts was always an object effentially neceffary to the *greateft captains.*

" It

"It is," says the commentator on Polybius, "one of the principal qualifications requisite in the commander of an army, and perhaps the least known." B I. Ch. 14.

I will add, that it is by the help of this science only, that an army can encamp with safety, that it may rest from fatigues, and screen itself from the continual inquietudes that the enemy's parties might give it.

It is now no longer a doubt, that war, like other arts, is to be studied, both in the closet and by exercise; a thousand examples have proved that an officer, who applies himself both these ways, has an infinite advantage over another who goes on in the vulgar tract, and learns only by rote.

"It is fine talk to say to an officer, be firm and courageous, never retreat, conquer or die; these maxims and rules, says M. de Botie in his treatise of Military Studies, make no impression on the heart of a man, but in proportion as his mind is enlightened, by knowing the methods "of

" of conquering, or blind to the dan-
" ger of being overcome."

In truth, it is study that opens our understanding, and excites our application; it is by that we supply our want of experience, by that we acquire those qualities which form great officers, and by that we open to ourselves the way to renown.

In the general operations of war, such as sieges and battles, the glory aspired after by all military men, is reserved for the superior officers only, because in these great actions every thing is attributed to them, and put to their account. Therefore it is only when a private officer, having the chief command of a party, can make a gallant defence, or can execute an enterprize to be talked of, that he may thereby be the instrument of his own glory, may merit the commendations of the army, and the favour of the court.

What satisfaction must a young soldier feel, when by various devices he so opposes his enemy, that he secures himself from surprizes, resists his attacks,

tacks, disconcerts his projects, and makes him abandon his enterprize! Comparable then to the greatest captains, he has a share in the prosperity of the arms of his sovereign, and in the defence of his country, and merits so much the more, as one of our masters in the science of war, asserts, " that the glory acquired in the de- Follard, " fence of a weak post, is infinitely Vol. 5. " above what may be gained in the " most important fortresses of the " state."

The means of acquiring this glory, never depends singly or merely on the greatness of one's courage, which is useful only in the execution; but on the combination of the talents that are necessary both to contrive and execute a project with success.

That bravery which elevates us above all dangers, is not sufficient; it may even turn to our damage, unless a wise and enlightened conduct restrains us from rushing thoughtless and rashly into action.

There will be no detail in this treatise, of what belongs to those posts, to which *general* officers are commonly detached; nor rules for the construction of lines to establish a communication, and ensure the safety of an army.

As I write only for *private* or *regimental* officers, I shall treat only of such posts, as they may be detached to, with thirty, fifty, or one hundred men; and will lay down, 1st, Some general notions that they should have of geometry, to be able to trace out entrenchments. 2dly, The different works for fortifying posts 3dly, How to augment their force on all occasions. 4thly, How an officer should be prepared to go on a detachment. 5thly, How he should march towards a post that he is detached to. 6thly, How he may establish himself therein. 7thly, What precautions he should take to prevent his being surprized therein. 8thly, What dispositions he should make to maintain himself there with vigour. 9thly, The manner of defending posts. 10thly, and lastly, How to to attack them either

ther by open force, or to carry them off by ſtratagems.

Such is the plan of this work, from whence, I do imagine, that inſtructions may be drawn, to fortify, defend, or attack, even the moſt conſiderable poſts, as well as the ſmalleſt; the rules in theſe reſpects being the ſame, and differing only in the ſize of the works, which muſt be proportioned to the number of men, that the party conſiſts of.

Whatever ſimilitude there ſeems to be between the ſervice in time of peace, and the ſervice in time of war; I will venture to ſay, that they will ſcarce bear a compariſon, and that they ſhould be rather conſidered as two different profeſſions.

In garriſons the ſervice is extremely ſimple, there being no more requiſite, than to know how to obey. In camp, it is connected with a thouſand accidents, that require an officer to have ſkill to command well, and act a determined part. Any one will be convinced of this difference, when they have ſeen my particular opinions, ſupported by

C   general

general examples. The relation of facts being of all methods of writing the moſt uſeful, moſt inſtructing, and moſt amuſing; I will quote both good and bad, analogous to each article: the latter, that we may learn wiſdom at the expence of thoſe who have lived before us; and the former, that by ſeeing the gallant actions that have been performed, we may be excited to imitate them.

There are no inſtances wherein military virtue ſhines with greater luſtre, than in thoſe where ſeemingly our weakneſs ſhould cauſe our defeat.

To tire out a ſuperior enemy, who expected to have led us off in triumph; to repel him, and throw upon him the ſhame of a broken and ill concerted project;—this is what characterizes the great officer; this is the higheſt ability a ſoldier can be maſter of.

Let it not be imagined that what I ſay here, are only high ſounding words void of ſenſe; the examples which I ſhall cite will prove their reality; and the means I ſhall propoſe, will prove

the

the facility of executing actions of this nature.

However little the affistance was that I could find in the military authors that I confulted, I have endeavoured to omit nothing in this work, that may ferve to make that part of war that I treat of, underftood; and I have taken much lefs pains to adorn it with wit, than to furnifh it with principles which may be relied upon, and rules that may be eafily followed.

I will explain what the principles and rules are, when I have given an idea of geometry theoretical and practical, which an officer muft neceffarily be mafter of, in order to know how to make an entrenchment.

[13]

THE
SCIENCE
OF
MILITARY POSTS.

CHAP. I.

Geometry necessary for an Officer.

GEOMETRY is the noblest part of mathematicks; it is the science of measuring all things that have perceptible dimensions.

Officers of infantry being never employed to direct the construction of

great

[14]

great fortifications, (that being only the bufinefs of engineers) the geometry, neceffary for them, may be reduced into a very fmall compafs.

The *field fortifications*, which they may probably direct, are fo fimple, that the only requifite knowledge is, to trace ftrait lines, horizontal, parallel, perpendicular, and curved ones, and to underftand the connexions between them, in order to execute them.

But as the operation of drawing thofe figures with compaffes and rule on paper, is very different from that of tracing them with the fathomed line on the ground, I will fay a few words, in order to explain them both.

Of the Point, and of the Line.

Plate I. Fig. 1.
Geometricians call (a Point) the fmalleft thing that can be imagined; it is confidered in mathematicks as indivifible; that is to fay, having no dimenfions.

A line drawn by a rule from one point to another, is called a right line,
as

as A, B; this line is the shortest that can be drawn between the same points, and is considered, like all lines in general, as a row of points placed side by side, in a strait or right line.

This line A, B, is called horizontal, when it is so level, as neither to rise nor fall towards A, or B.

C, is called a perpendicular line, Fig. 2. being drawn right up and down, so that it neither leans towards one side or the other, such as a thread would mark, having a ball suspended from it.

To draw a line D, perpendicular to Fig. 3. a strait line E, F, the point from whence the perpendicular is to be drawn, may be out of this line, or in the line itself.

If the point G is out of the line H, I, Fig. 4. then from this point, as center, describe an arch, which shall cut the line in two points, as L, M; from these points L, and M, and with the same distance, or radius, describe two arches intersecting each other in one point N, then draw a line from the given point G, through the intersection of the two arches N,
which

[16]

which line will be perpendicular to the line H, I.

Fig. 5.

But if the point O, from whence the perpendicular is to be raised, is in the line itself S, T, then from this point, as center, describe a semicircle which may cut the line in two points P, Q, from which two points, as centers, you will describe the arch R, with the same opening of the compasses, and draw from the point O, the line R, O, thro' the points of intersection of these arches, which will be perpendicular to the line S, T.

Fig. 6.

If, in the second case, the point from which you are to raise the perpendicular, was at the end of the line V, X, you must then prolong this line beyond the point V, to describe from this point, as center, a semicircle, which may cut the line in two points, and do the rest of the operation as above.

A line is *oblique*, when it leans one way or the other from a perpendicular.

Fig. 7.

The *tangent line* (b) is that which touches another line in one point only, without

without cutting it; it is different from a fecant line, which is that which cuts another line.

A *curved line* is that which differs from a ſtrait line, in going from a point A, to another point B. Fig. 8.

A *mixt line* is that which has part of it ſtrait, and the other curved, as C, D. Fig. 9.

A *ſpiral* is a curved line, turning round, and always widening its diſtance from the center, as E. Fig. 10.

Two lines F, G, H, I, are parallel, as they are equally diſtant from each other, ſo that they would never touch each other, though they were to be prolonged to infinity. Fig. 11.

If from a given point F, you would draw a parallel to the line H, I, deſcribe from this point F, and with a ſpace taken at diſcretion, the indefinite arch G, L, then from the point L, and and with the ſame opening of the compaſſes, deſcribe another arch F, H, then take on the firſt arch a part L, G, equal to F, H, and draw a line to paſs thro' the points F, G, which line will be parallel to the line H, I.

D   Section,

Section, or intersection, is the point through which two lines or arches pass to cut each other, as M, M.

Fig. 12.

A *circle* is a figure contained within a single line, from all parts of which the center is equally distant, as N,O,P; every circle is supposed to be divided into 360 parts, or degrees, so that the semicircle is understood to be 180 degrees, and a quarter of a circle is 90. This division by degrees serves to measure the angles.

Fig. 13.

The *circumference* is the crooked line that describes the circle N, P.

A strait line Q, R, drawn from one point of the circumference to the other, passing through the center, is called the *diameter* of the circle, because it divides it equally into two parts.

Fig. 14.

A *semidiameter* is a strait line drawn from the central point of a circle to a point of the circumference, as S; this line is also called a radius.

A strait line T, V, which divides the circle into two unequal parts, is called a *chord*, and the portion of the circle X, cut by this line, is called an arc or arch.

Of Angles.

An *angle* is a space bounded by two lines which meet at a point, as A. Fig. 15.

If a line B, is raised perpendicularly Fig. 16. on a strait line C, D, the angles that they make are right angles; their measure is a semicircle, that is to say, they have together 180 degrees, and each of them 90.

Two right angles are drawn in the same manner as you raise a perpendicular in the middle of a strait line, as aforesaid.

Likewise, to make one right angle Fig. 15. 4 A, the operation is the same as is directed to raise a perpendicular at the end of a strait line; its dimensions are ninety degrees, or the fourth part of a circle.

An angle is *acute*, when the lines that compose it approach near to each other, as E, and the angle contained Fig. 17. between them is less than 90 degrees.

An *obtuse angle*, is that which is made by two lines, which open from each
other

other to a greater diſtance than a right angle; its meaſure therefore is more than 90 degrees, as F.

Fig. 18.

The top of the angle is that part where the two lines meet in a point, that compoſe it, as G, H, I.

Fig 19.

To meaſure the contents of an angle of any kind, place one point of your compaſſes at the top H, and deſcribe with an indefinite diſtance taken with the other, the arch or portion of the circle L, M, the dimenſions of this arch which touches both lines of the angle are the dimenſions of the angle itſelf, let the opening of the compaſſes, with which this arch was deſcribed, be what it will; the reaſon of that is, that whatever the opening of the compaſſes was, the angle G, H, I, will always be the quarter of a circle. The ſame rule ſerves for all angles.

Fig. 20.

To make an angle N, equal to another angle O: from the angle O, as a center, with your compaſſes opened to any diſtance, deſcribe the arch P, Q; then, with the ſame opening of the compaſſes, and from N, as a center, deſcribe

[21]

scribe the indefinite arch R, S, then take the distance P, Q, the size of the given angle, and set it off from R to S, and draw through this point S, the line N, T; this line will form an angle equal to the given angle O. Fig. 21.

It is so necessary in practice to know how to make one angle equal to another, that the execution of fortification, and all parts of the mathematicks, would be impossible without it; but what I have said serves for the construction of works sufficient to fortify posts in the field.

Of Triangles.

A *triangle* is a figure bounded by three sides, which form three angles.

An *equilateral triangle*, is that which has its three sides, and its three angles equal, as A, B, C. In order to describe Fig. 22. this, draw a first line A, B, then from the point A, and with the distance A, B, describe the arch D; then from the point B, and with the same opening of the compasses, describe another arch E; then draw through the point of intersection

[22]

section (c) the lines A C, B C; the triangle formed by these lines will be equilateral.

Fig. 23.   A *triangle* is *right angled*, when it has one right angle, as F.

Fig. 24.   An *Isosceles triangle* is that which hath two sides, and two angles equal, as G, H.

Fig. 25.   A *Scalene triangle* is that whose three sides and three angles are unequal, as I, L, M.

     The three angles of any triangle are equal to two right angles, that is to say,

Fig. 26. the three arches N, O, P, described with the same opening of the compasses, make together 180 degrees.

     The *contents* of the *surface* of a *right angled triangle*, or of one that has a right

Fig. 27. angle, as R, is equal to half the product arising from its height multiplied by its base, or to the whole product of half its base, by its height; because it is half of the square R, S, which would have the same base and the same height.

Fig. 28.   The line T, which is drawn from the point of an angle, perpendicular to its base, is called the height of the triangle;

angle; this line will form two triangles, which will of courſe be right angled, as V, X: you find the contents as above-ſaid *.

The doctrine of meaſuring triangles is called *Trigonometry*; which is one of the nobleſt parts of Geometry, and is treated at large by many excellent authors.

Of Surfaces.

Though the knowledge of *ſurfaces* and of *ſolids* is uſeleſs in conſtructing ſmall field fortifications, nevertheleſs I will explain them, as being an eſſential part of the principles of geometry.

Surface, or ſuperficies, is a figure determined by many ſides ‡.

A *plain ſurface* is that, which is even like a looking-glaſs; it is *convex*, when it riſes in form of a globe; and *concave*, when it has any depth.

The

---

* An Equilateral, or Iſoſceles triangle, ſo divided, will make the two new triangles equal each to each; but in a Scalene ſo divided they will be unequal, though always rectangular.

‡ Or in other words, a Surface is that which has length and breadth, for a circular Surface is bounded by one line.

[24]

The line being considered only as a row of points, the surface is also considered as a row of lines placed beside each other.

Fig. 29. A surface as A, whose four sides are equal, and the four angles right, is called a *square*; the surface or area of this square, which is also called a rectangled square, is equal to the product of its base, multiplied by its height; that is to say, if it is four toises or fathoms long, and four toises broad, its surface will be sixteen square toises, or sixteen rectangled squares, each of which will be a toise in base, and a toise in height.

Fig. 30. The *parallelogram* or *oblong square* B, is a figure whose opposite sides are equal, and the angles right; its surface or *area* is measured like that of the rectangled square, that is to say, if it be nine toises long, and four toises broad, its surface will be thirty-six square toises.

Fig. 31. A *rhombus*, or lozenge C, is a figure of four equal sides, but which has two
opposite

opposite angles acute, and the two others obtuse.

The *trapezium* D is a figure, whose four sides and four angles are equal. Fig. 32.

The figure E is called a *polygon*, which has many angles, and above four sides: A *regular polygon* is that, which has all its sides and all its angles equal; by the term *polygon* is sometimes understood, the whole of a *fortified place*, and sometimes only the ground traced out in order to raise the works. Fig. 33.

A *polygon* that has five sides, as E, is called a *pentagon*; *hexagon* that which has six; *heptagon* that which has seven; *octagon* that which has eight; *enneagon* that which has nine; *decagon* that which has ten; *undecagon* that which has eleven; and *dodecagon* that which has twelve.

When a place is fortified with six *bastions*, it is said to form a *hexagon*. It is regular, if its sides and its angles are equal; and irregular if they are not.

[26]

All surfaces, except a *circle* *, that are not rectangled, are measured by dividing them into rectangled triangles; of which I said that the contents were, the product of half the base, multiplied by its height.

Of Solids.

A *solid* is a figure which has three dimensions, viz. length, breadth, and thickness.

As lines are only rows of points, and surfaces only rows of lines, in the same manner solids are only rows of surfaces, supposed to lie one on the top of the other, like the leaves of a book.

Of the different kinds of *solids*, the principal are, the *cube*, the *parallelopipede* or *parallelipede*, the *prism*, the *cylinder*, the *pyramid*, the *cone*, and the *sphere*.

Fig 34.
The *cube* is a figure, whose length, breadth, and thickness are equal, as F; a *die* is a cube. The contents of a cube

---

* The surface or area of a circle is found by multiplying the circumference by one quarter of the diameter. *See the following note on the sphere.*

[27]

cube are found, by multiplying the length by the breadth, and the product of those two by the thickness; that is to say, if it be four fathoms long, and four fathoms broad, whose product is sixteen square fathoms, in multiplying those sixteen by four of depth, you will have sixty-four cubic fathoms, or sixty-four solids, each of which will have one fathom in every one of their three dimensions. Plate II.

A *parallelipede* is a solid, bounded by six parallelogram sides, whose opposites are parallel and equal, as G. You find Fig. 1. the contents of a parallelipede, by multiplying its dimensions one by the other, like the cube.

The *prism* is a solid, which has an equal thickness in its whole length, and whose upper and lower bases are equal: this name is particularly given to a triangular solid, as H, bounded at its Fig. 2 two ends by two triangles, equal and parallel to each other, having three parallelograms for its sides, which cannot be parallel to each other.

E 2 The

[28]

Fig 3.
The *cylinder* is a round body, equally thick in its whole length, and whose bases are equal circles, as I; the contents of prisms and cylinders are equal to the product, arising from the area of their base, multiplied by their height.

Fig. 4.
A *pyramid*, as L, is a solid, whose base is square or triangular, and which ends in a point; a pyramid is the third part of a prism of the same base, and of the same height.

Fig 5.
A pyramidal figure, as M, is called a *cone*, whose base is a circle, and whose top ends in a point; its solid contents are found, like those of the pyramids, by multiplying the area of its base by one third of its perpendicular.

Fig. 6.
A solid that is round, as N, is called a *sphere*, like a globe, or a ball. The *measure* or *area* of the *surface* of a *sphere*, is the product found by multiplying the circumference * by the diameter;

* By knowing the diameter of a circle, the circumference is found by the rule of proportion. For as 7 is to 22, so is the diameter to the circumference; or if the circumference be given, then reverse the proposition, and say, as 22 is to 7, so is the circumference to the diameter.

meter; and the solid contents of a sphere or a globe, is the product found by multiplying the area of the surface by one third of its radius, *i. e.* one sixth part of the diameter.

Of *Practical Geometry*.

The rules of practical *geometry* are the same as those of theoretical geometry, whose object is to reduce the principles to use, and to mark on the ground, the different figures that may be drawn on paper.

The necessary instruments to trace geometrical figures on the ground are, a *toise* or *fathom*, a *chain*, a *plain table*, *picquets*, a *level*, and a *plomb*.*

In the place of the toise, which is a wooden measure of six feet long, each foot composed of twelve inches, &c. one may use a sword of three feet long, which will be found sufficient. This is a convenient length for a sword, and on all

---

* But as those may be too many for every officer to carry, I will shew how their places may be supplied on all ordinary occasions.

all detachments an officer will have his meafure with him.

In the place of the *chain*, which is made of iron, and divided into five, and five, or ten, and ten toifes, by a pendant; one may ufe a line divided into toifes by fo many knots. This fhould be fixty feet long, and a loop handle fhould be left at each end, the length of which is to be included in the firft, and in the laft toifes. This loop ferves to pafs on the picquet that is to be drove into the ground, from which the line is to be ftretched. It is true that this line will be fubject to fmall alterations by being wet or dry; but field forts, not intended to laft long, do not require a fcrupulous exactnefs.

The *plain table* is one of the feveral inftruments ufed to take angles in furveying, but too cumbrous for an officer to carry, nor indeed is it neceffary in what we intend to treat of here.

Picquets are fticks of about three feet long, and of an inch or an inch and a half diameter, fharpened at one end. They ferve to ftretch the line, by driving

ing them into the ground, with a mallet or great stone, to draw the lines, and to mark the tops of the angles of the intrenchments that are to be drawn: but as wood is to be found almost every where for this use, one need not carry them; but it is enough to know that twenty of them are necessary for a square redoubt.

As the *level* may be omitted, I shall only say that it serves as a guide to make the earth even, and to draw horizontal lines.

The *plomb* serves to draw perpendicular lines in carpenters and masons work; but it may be supplied by tying a bullet or a stone to the end of a string, which you may hold before you to the work.

Such are the instruments used in tracing on the ground; those which are used after to dig the earth, are *shovels* and *pick-axes*; and to cut wood, *hatchets* and *bill-hooks*, all which are absolutely necessary, and it is impossible to substitute any thing in their places. Therefore a body of men ought never to march to a post, without being provided

vided with one or two of each kind. I now come to the method of tracing.

Fig. 7.   To trace a ſtrait line from A to B, I plant a picquet at the point A, on which I fix the loop which is at the end of the line, which I draw tight towards B, with a ſecond picquet: having done this, I mark with the point of a third picquet, a track on the ground, gently touching the line all along as I mark.

Fig. 8.   To raiſe a perpendicular line at the point C, of the line D, E, you muſt fix a picquet at the point C as center, on which you muſt hang one loop of the line, and deſcribe, in turning, with the point of another picquet (alſo faſtened to the line, at the diſtance of a toiſe) the ſemi-circle F, G, then from thoſe points F and G, where picquets are to be drove, you ſhould trace with the diſtance of two or three toiſes, the arches H, I, then fix the line again to the picquet C, which draw tight towards L, in paſſing thro' the ſection of the arches H, I, and draw the line C, L, which will be perpendicular to the line D, E.

If

[33]

If this perpendicular M, should be raised at one of the ends of the line N, O, you must prolong this line towards P, trace the semicircle Q, R, and finish as before said. Fig. 9.

But if the point S, from which a perpendicular line is to be drawn, was out of the line T, V, you must fix a picquet at this point S, on which you must fasten your line by one of the loops, and trace the portion or segment of a circle intersecting the line T, V, in the points X, Y, then from these points X and Y, where you must fix picquets, you must trace with the line, with an equal distance, the arches Z, and make the line S, Z, pass thro' the point of intersection Z, and the given point S, which will be perpendicular to T, V. Such is the method to trace a perpendicular with precision; but as this operation is a little complex, and as one may not always have time to execute it, I believe it will be enough to see that the cord when stretched from *a* to *b*, in order to trace the perpendicular required, forms, Fig. 10.

Fig. 11.

F

[34]

as near as the eye can judge, two right angles with the line *c, d*.

Fig. 12. To trace two lines parallel, you must after having drawn the first line *c, b*, measure the distance to which you would trace the parallel, which I suppose to *g*, describe from this point *g*, as center, (where you must fix a picquet) the indefinite arch *h, i*, and from the point *h*, describe the arch *e, g*; then take on the first arch the part *h, i*, equal to *e, g*; and lastly, trace the line *g, i*, this second line will be parallel to the line *e, h*.

But as these works, which officers may have occasion to make, are of but little extent, and that great nicety is not required, one may shorten this operation, by measuring towards the two ends of the line *e, h*, with a sword of the length before-mentioned, the two equal distances *e, g*, and *h, i*, as perpendicular as it is possible to judge by the eye, and trace afterwards thro' the marked points the line *g, i*, which will be parallel to the line *e, h*.

When I shew how to trace square redoubts, I will explain the ways to trace a triangle, a perfect square, and a circular figure; but I will not speak of regular or irregular polygons, because those figures are only used in the construction of great fortifications.

As the operations which require the tracing strait lines, perpendiculars, paralels, and angles, are those which are oftenest necessary in field fortifications, young officers ought to practise them often.

There is often a great deal of leisure time in a settled camp, as well as in garrison, in time of peace, which might be employed in those amusements, where willing soldiers may be made to assist, by giving them a small recompence; I say by soldiers rather than peasants, because this advantage will accrue; that is, they will work cheaper, and it will accustom them to work with more ease when the war may require their service.

CHAP.

CHAP. II.

Of the different Works with which Posts may be fortified.

IT may be expected that I should speak of the *detachments* to *posts* before the manner of intrenching them; the detachments being always ordered to work thereon; but the drawing geometrical figures, having led me towards *intrenchments*, I think them proper to follow.

The security of an army depends on the *defense* of its *posts*, and on the vigilance of the detached guards.

Let the abilities of the General be ever so great, it is impossible that he can have an eye to all the little details that contribute to their defence; it is sufficient if he knows that the guards are well posted, and that the line they form be well supported. It is afterwards the duty of the several officers that command them, to make the best dispositions for a vigorous defence, that they may answer the General's views.

An officer is detached to a post, either to relieve a party, or to take first possession of it himself: In the first case, it often happens that the guard to be relieved is intrenched; as soon as he arrives at the post, and has taken the charge thereof from the officer that commanded, he is to prepare for his defence, as I shall explain on this article. In the second case, if an officer that is detached would intrench himself, he must observe, first, to chuse the place, to throw up his intrenchment, so that he may from thence discover all approaches; for if the enemy could come, without being perceived, to within a small distance of his post, the assailant might cover himself and his party, where he might remain in safety, and keep the besieged always under arms, taking time to advance upon them whenever he pleased

If on the ground, where he would throw up his works, there may be any hollow ways, a thicket of wood, or any other

other means of the enemy's being covered, they muſt be rendered uſeleſs to the enemy, or guarded by detachments of ſix or ſeven men. Secondly, He ſhould take care that there ſhould be no high place near that may command him; or he muſt hinder the enemy to profit by it: becauſe if the enemy ſhould take him in flank, or fire into the work, it would be impoſſible for the ſoldiers to defend it. The manner how to avoid this inconvenience ſhall be explained hereafter, when from the nature of the ground there is no taking poſts to avoid thoſe heights. Thirdly, He muſt make his work in proportion to the number of his men who are to defend it.

Good ſenſe and various examples ſhew that too large intrenchments, ſuch as are frequently made, cannot be well defended but by a conſiderable number of men. An exceſs of this kind ſeems to me a great fault. I think it better to fall into the oppoſite, in making them ſmaller. Fourthly, He muſt be attentive to give equal ſtrength to
all

all parts of the work, that it may resist the enemy alike every where. Fifthly, and lastly, He must try strictly to fulfill the intended schemes of the General, by maintaining his post.

If he is detached to an open countay, or to a height, and that he may be attacked on every side, as is often the case with small guards, he must build a redoubt, or small square fort, with a parapet, a banquet, and a ditch.

When the ground is chosen, he should mark a strait line a, A, E, and raise the perpendicular a, B, C, as described in the article of Practical Geometry; set off from a towards c, and from a towards E, the dimensions proposed for each side of the parapet within the fort, which, if his party consists of thirty men, should be two fathoms or two and a half, four fathoms for fifty, and eight fathoms for one hundred men: this will afford a space opposite the parapet of almost two feet for each man This space no doubt will appear too large for the defence of an intrenchment, where the men are to be

Fig. 13.

drawn

drawn up at least two deep; but it is impossible, neither is it ever the practice, to make their proportions less, except only when the detachments are very numerous; the size of the parapet may be so proportioned, as to admit the men who are to line it to be drawn up two and even three deep.

After having drawn those two first lines A and B, as before said, the loop of the cord is to be fixed or hung on the picquet C, of the perpendicular line B, and with the same length a, c, trace the arch D; then hang the cord on the picquet E, at the end of the line A, and with the aforesaid distance, or length, describe the arch F; the point of intersection in those arches determines the length of the lines E, H, and C, G; those four lines so drawn form a square, which will mark the interior sides of the parapet.

Then four other lines I, L, M, N, are to be drawn at the distance of two or three feet within side this square, and parallel to the first, to mark the breadth of the banquet, which is to be

more

more or less, according to the intention of drawing up the men, who are to stand thereon, in more or fewer files.

In the next place, a third parallel square, O, P, Q, R, is to be drawn without-side the first, to determine the exterior side of the parapet, and its thickness, which is commonly eight or nine feet; but if the intention is to resist cannon, it must be eighteen feet, or three fathoms.

Lastly, a fourth square, S, T, V, X, must be drawn to determine the breadth of the ditch, which must be the same with the parapet, or may be two feet more than the thickness of the parapet, and a picquet must be fixed at every one of the angles, as well as at those of the lines already marked, to avoid losing the points of the square.

While the drawing and marking are executing, with the help of two or three men, five or six more should be employed in cutting down the nearest trees to the post, as well to discover the approaches of the enemy, as to serve in making the intrenchments: the small branches serve for fascines, which are a kind of faggots about a fathom long, and two feet thick, uniform from end to end; they are bound at each end, and in the middle, and serve in the intrenchments to keep up the earth, which, without them, would crumble down; the middling branches of these trees serve to make stakes, or picquets, to drive between, or through the fascines, to fasten them to the ground, or one on the top of the other, in order to raise the parapet. Lastly, the trunks and greatest branches are used to strengthen the post, as shall be shewn hereafter.

The drawing being finished in the above manner, the first range of fascines is to be laid on the smallest square, I, L, M, N, as a foundation to support the earth of the *banquette*; then the second range is to be laid on the square A, B, G, H, to keep up the inside of the parapet; then the third range on the square O, P, Q, R, to keep up the outside of the same parapet.

You

You must remember, when you are pinning down, or picquetting your first fascines, to leave on the side least exposed to the enemy, a space of three feet, for entrance into the redoubt; but if it be in the power of the enemy to get round, so as to fire right thro' or *enfilade* this passage, then let it be made winding like the figure Y, in which form it Fig. 14. cannot be enfiladed.

After having picquetted the three ranges with fascines, as I said before, you must dig the ditch A, B, distant a Fig 15. foot from the outside of the parapet; this space, or breadth, which is called *berme*, serves to keep up the mould, and to receive the rubbish thrown down by the enemy's cannon from the parapet: this *berme* is broader, or narrower, according to the stiffness of the earth; throw the mould into the intervals C, D, E, marked for the parapet and the *banquette*, and make the men tread down the earth, so as to make it hard; also take care, in hollowing the ditch, to leave a *talus*, or slope, more or less, according as the nature of the ground requires, to both sides F, G, to prevent the earth from falling down.

The slope F, which is next the redoubt, is called the *scarpe*; and that on the outside next the field, as G, is called the *counterscarpe*: You must take care in picquetting the fascines, as you raise the parapet, to bring the fascines of each face a little closer together, as you see at H, so as to leave the same slope of each side the parapet. The space E, D, marks the *banquette*, the space D, C, the thickness of the parapet below, the space I, L, the thickness of the same parapet at the top; the space M, N, the breadth of the bottom of the ditch; and the space A, B, the breadth at the top of the ditch.

You are to raise the *banquette* of this work two feet, if the ground is even; but if there are some places too low, you must make two *banquettes* one over the other with steps. You must make the parapet four feet higher than the *banquette*; but if this *banquette* was risen on account of some neighbouring height,

height, from whence you might be enfiladed, you must also raise the parapet, till you see the enemy cannot hurt you.

Leave to the upper part of the parapet a *talus*, or slope, I, L, that your men may see all without, and to fire well towards the country O.

Tho' the form of a square redoubt, as I have described, may be what is almost always used in the field, it has nevertheless defects, which ought to be an objection to its use, at least on posts that are to be defended on all sides alike.

Experience shews us, that the defence, by oblique firing of musketry, is not to be depended on; for a soldier hardly ever fires in any other direction than right forward, as A, and even without aiming. By this way of firing Plate II. it happens, that great spaces, opposite Fig. 16. the angles of the redoubt, as B, are left without defence, where the enemy may remain in safety.

The Chevalier de Clairac, an experienced and good engineer, gives, in his
treatise

[46]

treatise of slight or temporary fortifications, an excellent way to remedy this inconvenience; it is to make the inner side of your parapet indented, or as it were a row of little redans, fit for one or two men of a side: this method is Fig. 18. the more excellent, as the cross-firing takes the enemy on each flank, and there can be no approach that is not defended. But this kind of redoubt has too much work in it, and takes up too much time in constructing, to be made by regimental officers; I would join with the same author, to make cir‑
Fig. 17. cular redoubts C, because every point of the circumference being equally disposed, the soldiers may stand any where, and the exterior spaces D, which are defended, changing every minute, the enemy has no place of safety.

The circular redoubt therefore is the most perfect that can be made; but when you have a road, or the bank of a river to defend, the square redoubt, the oblong, or triangular, are preferable; because you ought to oppose the faces of your intrenchment as low as
possible,

[47]

possible, and parallel to the parts that you would attack, always observing to round off your angles.

In order to draw a circular redoubt, after having chosen your center, you must fix a picquet at the center point; and from this point, with a determined length of line, according to the number of your people*, describe the circle E, E, to mark the inside of the para- Fig. 17. pet; then draw another within that, allowing the breadth that I said before for the *banquette*; then draw a third F, F, to mark that of the parapet; lastly, draw a fourth G, G, to give the outside breath of the ditch : this done, picquet your fascines, in giving them the curve line of the circles, and finish as in the square redoubt.

If

---

* If you have thirty men, give a fathom and a half to your cord, for the length of the radius, which will make three fathom diameter, and about nine fathom circumference, and a little less than two feet for each soldier. If you have fifty, you give two fathom radius, which makes four fathoms diameter, and about twelve fathoms round. Lastly, if you have one hundred, you double these proportions, unless you mean to draw up your men two or three deep.

If an officer commanded a detachment that was posted on the pass before a bridge, in a defile, or before a ford, he might make a parapet in a strait or a curve line with its *banquette*, and a ditch that should stop all the entrance: this would be better than a redan, which is a work that has but two faces. I will not describe this work here, because one officer never has men enough to defend the parapet, which is commonly very extensive. For the same reason I will not speak of star forts, or the larger forts, where only great detachments will do.

I will describe the ways to fortify houses or villages, when I come to shew the manner of adding fortifications to such posts as are already strong by nature.

C H A P.

CHAP. III.

Of the different ways of encreasing the strength of Posts.

IT is not only with the works that I spoke of, in the former chapter, that an officer may fortify his post; there are also an infinite number of ways how to stop, to tire out, and even to repulse the enemy, of which he should not be ignorant.

The strength of a redoubt A, or of any other work, may be encreased by filling the ditch B with water, which is done by turning the stream of a spring or rivulet into it, or by cutting a drain from a river or pond. But if the ground of your post is uneven, which sometimes happens, and that you cannot run the water equally into all parts of the ditch; you must, in hollowing it, make *batardeaux* or dams, as C, or little traverses of earth, which make proper banks to keep up the water in the higher parts of the ditch D, from whence

Plate III.

you may let it run down into the lower, E. You muſt allow to theſe banks at the top D, only half a foot thickneſs, which will be pretty ſharp; but you muſt leave it much thicker at the bottom E, giving a great ſlope to each ſide.

Dams, like F, are alſo made with boards or planks; but they muſt be ſtrong, and kept up by great ſtakes, ſo that the weight of water above ſhould not puſh them down: This kind of dike is preferable to thoſe of earth.

You may alſo ſtrengthen your works, by ſtopping up or embarraſſing the environs and the avenues that lead towards you. In a mountainous country you may interſect the roads with large ditches, or break them ſloping; you may ſtop up the defiles with waggons one a-top of the other, and poſt a few muſketteers behind them; laſtly, you may * lay heaps of briars and thorns half buried in the ground, wherever you

---

* I think if roots of trees, or large branches or blocks of wood, are placed, part of them buried, they will do better.

you find it neceſſary. But if, as it often happens, the General orders the officer to retreat towards the army, or to fall back to another poſt, if he ſhould be attacked; he muſt take care not to deſtroy the road ſo as to hinder his own retreat, but muſt leave himſelf a paſſage, made like a draw-bridge, or ſome other way, which muſt be guarded by ſeven or eight men.

If he is detached to a plain or flat country, he muſt dig deep ditches in the avenues to it, and in the approaches to the poſt; or pits, which he may cover afterwards with ſlight boughs, and a little earth over them, and take care to ſpread the earth dug out of theſe pits on each ſide, that the enemy may not know exactly where they are. Alſo he may ſcatter in the avenues caltrops, which are, as it were, iron ſtars with four ſpikes, ſo diſpoſed, that whatever way they are thrown, one ſpike will always ſtand pointing upwards. Laſtly, he may fix picquets quite round his poſt, very near each other, and a little inclining to the field outwards, about two feet out of the ground, and sharpen their points afterwards.

But the greatest obstacle that he can oppose, is what M. le Chevalier Follard depends most on, in his Commentaries on Polybius; which is, to shut up the roads, and embarrass the defiles, and surround your post with a breast-work of trees, with their trunks buried about three or four feet in a ditch made on purpose. The trees fit for this use ought to have large branches, and you may sharpen their points, and take off all the leaves; place the trees as near each other as you can, so that the branches may twist into one another, and see that they point a little towards the enemy. You may make, if you chuse, three or four ranges of breast-works of trees round your redoubt; but they must be at two fathoms from each other, that the enemy may not burn them at the same time, to approach your redoubt.

" Good redoubts," says M. de Saxe, in his Reveries*, " are the most ad-
" vanta-

* The pocket edition of Reveries of M. de Saxe, pag. 326.

" vantageous, as they are the soonest
" made, and they serve on very many
" occasions, where one only, in a close
" country, will sometimes stop a whole
" army from annoying you on a critical
" march, and enable you to occupy a
" deal of ground with a few troops."

Sometimes you may cover your party by a simple *abatis* (or breast-work of trees) when you do not intend, or have not time to throw up one of earth; and you must take care to place the trunks one on the other, as much as you can, to make a kind of close parapet; otherwise the enemy, by forcing their way to the breast-work, and having openings to see even from the feet to the heads of those within the work, may kill them one after another.

If it is a ford, or a river, that you have to defend, you may make a parapet, minding to make it as near the water as you can, so that the enemy may not have ground to draw up, when they have passed over. You may add to the difficulty of the passage, in digging a deep ditch before the ford, and letting
the

the river-water into it; you may also slope down the bank sides, throw trees across each other, and lay caltrops. But all these devices, I have shewn, only serve to encrease the force of the outskirts of your post, which are next the enemy; there are other means that you may use besides what I have spoken of, in places of some natural strength, as *chateaux*, chapels, farm-houses, or barns.

An officer who is sent to a post of this kind, that stands by itself, must take care, before he begins to work, to oblige the inhabitants to quit it, and lodge them, by the help of the magistrate of the next village, in some other place. The next thing to be done is, to make a curving parapet, if he has people enough to defend it; if not, and if he has but a few men, he must make a breast-work round the house with trees, and chiefly before the angles, to prevent the enemy from undermining it there. He must also take off the tiles or slates from the roof of the house, so that the enemy, by putting ladders against the walls, may

not

not knock thofe within on the head. If the houfe is thatched, or covered with any other matter that is combuftible, it muft be taken off, and burned, left the enemy fhould employ it to demolifh the houfe itfelf; and he muft deftroy all other things of the like nature, for the fame reafon. This is the advice of M. Follard.

Though you have furrounded the houfe with a parapet or *abatis* (i. e. felled trees) you muft break fmall loop holes through the wall of the ground floor, and let them reach to within a foot of the ground, fo as to difcover the enemies legs, and to hinder them from taking your outworks, by your placing five or fix mufquets therein.

Thefe holes, which may be about four inches wide, ought to be broke through at about three feet diftance from each other; and there fhould be a little trench dug, at a foot and a half from the wall, and within-fide the houfe, wherein the men fhould be placed to defend it.

Also make other holes, seven or eight feet from the ground, oppofite the intervals of thofe below, and of the fame fize; and let your foldiers that are to defend thefe ftand on benches made of boards, or planks, or tables, or ladders; and take care to make a great number of holes oppofite the avenues leading to the doors, or the angles of the houfe; becaufe here you may expect the ftrongeft efforts of the enemy. If there is a court to the houfe, you muft make holes through the walls that look into it, fo as to be able to warm the enemy, if they fhould get into it.

If there are many doors, you muft barricade them, and ftop the paffages towards them, by laying four or five trees one on the top of the other, leaving to the one that is to be the entrance of your poft, only room for one man at at a time to come in. If there are low windows, that are not defended with iron bars, you muft ftop them up with dung, planks, ftones, earth, and even trees.

If

If the houſe has large offices on the ground floor, or the like, you muſt bury trunks of trees in the middle of them, leaving the branches ſticking up, to hinder the enemy from forming, if they ſhould get into the houſe. Laſtly, you muſt be attentive to plant one or two of theſe trees three or four feet within the entrance port or door, to hinder the enemy from forcing in front-wiſe, and to lay them under the difficulty of ſqueezing in ſide-ways

If there are great ſtairs in the houſe to go up to the firſt ſtory, you muſt break them down, or ſtop them up with ſtones, or caſks filled with earth. If the ſtairs are in a wing projecting from the houſe, as it often happens, you muſt break holes through the walls, to fire on the enemy that ſhall have got in; and you muſt uſe ladders to get up into the firſt ſtory. Alſo make ſeveral holes, through the floor, of four inches diameter, to fire from above on the enemy below. You muſt not make theſe holes in the boards over the place where you have fixed the trees, but

I make

make many over the door, and at all the weakest parts, where the enemy are likely to force in.

You may also break holes through the walls of the first story, about three feet from the floor, and leave the holes at least ten inches wide, and let them be broke three feet asunder, and over the intervals of those of the lower story.

As to the windows of this story, if you have not people enough to defend them, you must stop them up, to hinder the enemy from placing ladders, to fire in on you. You may, says the Chevalier Follard, make a great opening, before each window, in the floor, somewhat broader than the window, which will be a kind of ditch, into which those that should attempt the window must tumble.

You may make the same preparation in the second and third stories as in the first, that if the enemy would undermine you below, or break in above, they may find an equal resistance every where; but it will be useless to

make

make holes in the top ſtory, as you may, (having taken off the roof) take down the wall to breaſt high, and fire over it; reſerve the ſtones or bricks, and place them in heaps, to throw down on the enemy, and the rafters, to beat down the ladders that may be applied to the wall.

A poſt intrenched in this manner, may hold out a long time, and even tire out the beſiegers, if it be defended by reſolute ſoldiers; and ſuch are always to be found in an army.

M. D'Enfernay, who was very expert in fortifying poſts, ſuch as I mentioned, was detached, in the campaign of 1748, to Bevera, a village on the weſt ſide of Genoa, two leagues from Vintimillia on the Roya, with a free company, which he commanded: he took poſt in the church of the place, which ſtood by itſelf, and encompaſſed it with a parapet and wet ditch; but part of his intrenchment was commanded by ſome houſes of the village, ſo that the enemy might fire down on his people, and take them acroſs the parapet. He remedied this defect, in covering the part that was commanded with a kind of blind made of rafters, one end leaning again the church, and the other end fupported by pofts a foot higher than the top of the parapet, which gave his men liberty to fire under it; and it being covered with fafcines and earth, the fhot of the enemy could not hurt them within, who could however pelt thofe without.

I was detached from the army at this time with a body, to act under the orders of this experienced partifan; and I could not help admiring the defences that he contrived to his poft, where the enemy did not dare to vifit him, tho' they were his neareft neighbours.

I thought it my duty to mention this example, as well to do juftice to that officer, as to fhew the method of covering yourfelf in a poft that is commanded by a height. As for thofe that have no natural ftrength, fuch as redoubts, and other intrenchments of earth, this defect is remedied by raifing the fide of the parapet that is commanded

manded, as I said before, or in making a screen with rafters or poles fixed perpendicularly against the inside edge of the parapet, to which you may nail afterwards some planks, or fascines, observing to leave a space of half a foot between the top of the parapet and the bottom of the blind, for the musquetry to fire through.

But if an officer has not time for all those works that I have spoke of; which happens when a General has a mind to forage, and throws some foot into the houses, or farms, to form a line; an officer ought then to lay directly two trees across before the door, and cut holes through the floors, stop up the windows, and prepare for a vigorous defence, which will give time to the foragers to retire, or to detachments to arrive to succour them.

What I have said only relates to posts that stand alone; but if an officer should have a village to defend, he may cut out a more difficult work for the enemy: when I speak of intrenching a village, I mean only to speak of those where
the

[62]

the houses stand very close together, or of such as are sometimes enclosed with walls. A commandant sent to a post of this kind, ought, before he begins to work, to go several times round it, to examine the approaches, and the houses that are near it. Then he must make holes in the walls of some, as before said, and must block up the fronts of all such as have passages leading to the fields, with *abbatis* or felled trees; and if he has time, he may make a good *abbatis* quite round them, and intrench the entrance of the streets.

An officer that has a mind to fortify a post of some extent, in this manner, should sketch out a plan of the village, and of the intrenchments he intends to make, because that will give him hints how it may be defended, which, on the view only of the country, might escape him.

A street is to be defended like a bridge; that is to say, with a *redan*, or rather with a semicircular simple parapet, with a ditch. Break loop-holes in every story of the houses, as before described, which are near the entrance; make deep

trenches

trenches or cuts acrofs the ftreets, and lay caltrops in the bottom; alfo obftruct the ftreets with trees, carriages, or cafks; likewife open many paffages from the backs of the houfes, to go from one ftreet to another; and take care, efpecially if you have few troops, to fill up the middle of every wide place in the village with many trees, fo as to hinder the enemy from having room to draw up there, if they fhould get in.

As cannon or fire are the things moft to be dreaded in the defence of a village, an officer ought to break up the roads, to hinder their arrival; which is very eafy to do in a mountainous country; but when the village is in a plain, he muft cut deep trenches here and there athwart the avenues, and lay feveral trees acrofs, fo as to cover the whole breadth. If he has time, he may prop up the great beams of the houfes in the out-fkirts, with trunks of trees, or large pieces of wood put up like ftones: this is proper, left the floor fhould fall down by a cannonade, and crufh the men below.

In

In regard to fire, he had beft burn all combuftible matter, to prevent the enemy's turning it to his deftruction; but if there fhould be a great quantity of wood, ftraw, or hay, he fhould firft afk the confent of the General thereto; or, if he judges it neceffary, to have it carried off for the ufe of the army.

Another effential thing for an officer, who is detached to a village, to be attentive to, is to fecure himfelf a laft retreat, in cafe he fhould be forced in the ftreets, and at his firft intrenchments. For this purpofe he fhould chufe the *chateau*, (or *manfion-houfe*) the church, or any good houfe that ftands by itfelf, which fhould be intrenched with care, after having diflodged the inhabitants.

When foldiers who defend a poft know that they have a ftrong-hold to retreat to, they never think of giving up, or furrendering themfelves prifoners, while they fee they are in a condition to obtain honourable terms of capitulation.

But

But if the houses of the village to be defended, are scattered, or if there are gardens or courts in the center, you must then confine yourself to intrench a single house, or the church, or church-yard, or the *chateau*, where you may cover yourselves by an intrenchment of earth, and with all the other little artifices, that I mentioned for places that stand alone, observing always to take care of your flanks.

As the different works to fortify a post, and especially a village, are too troublesome, and even too tedious, if they were only to be made by the soldiers of a detachment, who ought never to be over-fatigued; an officer ought to command (with the assistance of the magistrate of the place) a certain number of peasants to work alone, or together, with a third part of the detachment, who should all be provided with shovels and pick-axes; while the other two thirds should remain under arms for fear of a surprize. These soldiers and peasants who work, should be relieved every three hours, and care should be taken

[66]

taken to see that they have no concealed arms, and that they continue their labour, without interruption, till the end of the work.

During the war in Italy 1747, I employed the inhabitants of the village of Berra, in the county of Nice, in this manner, where I was detached with thirty men. M. de Mirepoix, General of the French army, in the absence of Marischal Belisle, receiving advice, that some Piedmontese peasants had penetrated into this county, with a design to carry off some of our quarters; this General gave orders to all officers, who commanded the quarters, to be alert, and to send detachments to drive them back. In consequence whereof, M. de Charle, who commanded at the village of Contes, and in the district of Berra, wrote to me to put this post, one side of the court of which was quite open, into the best condition of defence that I could. The same day a spy of M. de Mirepoix came to my post, and told me, that two companies of one hundred men each, who were marching towards Berra,

Berra, would arrive there before midnight. On so exact advice, I ordered, by means of the magistrate, thirty peasants, who broke loop-holes in the walls of the *chateau*, wherever I directed them, and raised a good parapet of dry stones at the opening of the court yard, where the enemy might have entered by (ladders, or) *escalade*. Then I made them throw several trees, with their branches on, before this parapet; and sending for the magistrate and his eldest son, a lad of about fourteen years old, I told the father, that being informed by the orders I had received, that the enemy, who were marching towards my post, held intelligence with the inhabitants of the village, in order to carry me off, I would keep his son with me, to fix him on the top of the parapet; and that the first shot the enemy fired, should be at him. Having taken these precautions, I sent the father back, and held myself ready to receive the Barbetts; who hearing that we had parties abroad to attack them, fell back to Tende.

After describing the manner of strengthening a post, it follows in order, to speak of the detachments that are to defend them.

CHAP. IV.

Of the necessary preparations to go on detachment.

DEtachments are particular bodies of soldiers, sent from a greater body, to secure a post, or to go on some expedition.

When an officer is commanded to go on detachment, he should put on his gueters, sash,* and gorget, and take his fusie or spontoon, and provide himself with a line, divided into fathoms, to measure out his intrenchments, if he finds them necessary to be made.

He must be early on the parade or place of rendezvous. When he is come, and is told what party he is to command,

* English officers also take their sash.

mand, he should ask the Brigade-Major, whether he is designed to relieve another detachment? Or if he is to be the first to take possession of the post? If he is to relieve a body, he is only to require a guide to conduct him. This guide is a soldier sent by the officer to be relieved; who goes as an orderly man to the Adjutant-General, to be ready to carry any orders that may be necessary; and who having been at the post before, comes to conduct the new detachment. In case the post is to be taken for the first time, the officer is to ask the Brigade-Major for instructions relative to its defence.

Having taken these instructions, he must examine the men of his party, and be careful that every soldier is properly equipped; to see that his piece is loaded, and fresh primed, that his flint * is good,
that

* The translator has observed the English flints sometimes to be very good, and at other times very bad; and what he has seen of the French flints were uniformly good. He thinks a soldier ought always to be provided with a spare flint, or more. This inattention in regard to flints in a
British

that their stoppers are taken out, that their pouches are filled with ammunition, and that the rest of their accoutrements are in good order; and see that each soldier has his canteen and his bread for twenty-four hours, in which time they are commonly relieved: and never let a man be absent, either to eat or drink till relieved; and take care that his men carry proper tools to intrench with; such as shovels, bill-hooks, hatchets, and pick-axes, one or two of each sort; and if he wants any thing, he must apply to the Major of the regiment.

Some young officers may say, perhaps, that these are precautions that nobody thinks of taking; but are those then that I have shewn, so little essential as to be neglected? And is it not reasonable to think, if a soldier is unprovided of these things, that he will consequentially

British army, was very remarkable at the siege of Louisbourg; where it was a very usual phrase among the common soldiers in a skirmish, to say, " Ha! Monsieur, if our *flints* were as good as " yours, we should soon make you slacken your " fire."

sequentially be incapable of making the defence that he ought? It is to no purpose to say, that soldiers are of course supplied with every thing necessary of this kind in the field; for, on the contrary, I have often seen them wanting entirely every thing, so as to become useless members, and rather a burthen in a post.

"This, says M. de Vauban, is what
"makes us so unsuccesful in defence;
"owing to the neglect of many officers,
"in the provision of tools necessary on
"these expeditions: and the source of
"this neglect, which is too common,
"besides their ignorance and impru-
"dence, is, that they treat it as too
"trifling an article to merit their atten-
"tion; tho', in effect, it is one of the
"most important things to be ob-
"served."

As to war, when you form a plan of a good defence, it is better to take a thousand useless precautions, than to neglect one good one; because the least neglect may disconcert the best measures. But you must never be disheartened

heartened by imagining the enemy more vigilant than they really are, and by starting difficulties that most probably never will happen.* If in war you stop at every supposition that the imagination suggests, you will neither undertake nor execute any thing. One general rule in military projects is, never to forget any thing that may make us sure of success, and whereof the execution depends on ourselves only: as for what depends on the enemy, some part must be left to chance.

When the officer has examined his party, he may ask his guide the nature of the roads; if they be narrow, wide, open, or woody; if the enemy's posts are near; if they send out patroles; if he sees their parties in the country in the daytime: lastly, if he is to pass before any cottages, farm-houses, or *chateaus*; on these informations, the officer takes the necessary precautions for his march from camp.

CHAP.

* See the accounts of the expedition to Rochefort in 1757, and that to Louisbourg in 1758.

CHAP. V.

Of the march of detachments to posts.

THE march of a body of men to a post, is one of the most important duties on which an officer can be employed in time of war.

I will not say here, that night is the properest time to perform this march, because the time of their departure is fixed by the General himself: I will only observe, that there are certain necessary precautions, which ought never to be neglected.

If the post, to which the detachment marches, is distant from the camp, the officers should not get on horseback till they are out of sight of it, and should dismount when they come in view of the post, and have the horses led back by their servants; but if the place to which they are detached, is but a league, or thereabouts, from the army, and near the enemy, I think they had best go on foot, in order to be less embarrassed,

barraffed, in cafe of any fkirmifhes on their march.

But whether they go on foot or on horfeback, they, as well as their ferjeants, fhould take great care not to hurry the men too faft, left fome fhould not be able to keep up; to march clofe, and in as many files as the roads will allow; never to ftop, and to be very filent, fo as to hear all orders that fhall be given. You may fee in a treatife attributed to Marfhal Saxe, entitled, Traite des Legions, ou Memoires fur l'Infanterie, printed in 1753, of what confequence it is to a whole army, and to an officer particularly, to march in good order. The paffage is as follows:

" All the armies that the King fent
" into Bohemia, Weftphalia, and Ba-
" varia, marched off well equipped, in
" fine order, and very compleat; they
" returned ruined, worn out, and loft
" a vaft number of officers and foldiers;
" neverthelefs we had no confiderable
" action: the only one which was of
" any confequence, was to our advan-
" tage,

"tage. It was not by any visible
"stroke, but insensibly, that our army
"wasted away.

"In effect, the greatest part of the
"detachments sent out to skirmish, and
"posts that were distant, and escorts
"that were attacked by the enemy,
"were either surprized or beaten for
"want of discipline in the soldiery, or
"by the neglect of the officers. The
"man is not yet born that ever saw an
"escort march in good order: the
"soldiers continually employed in pil-
"laging, and stealing away out of sight
"of the officer, get the habit of strag-
"gling from the beginning of the
"march, and you'll scarce ever find an
"officer who gives the least attention to
"this abuse. It is the same in posts,
"parties, and detachments; either the
"soldier straggles, or if he stays with
"his party, it is only to march in bad
"order, stopping every minute, talk-
"ing when he ought to be silent, or
"murmuring when he ought to obey.
"If the enemy appears, they are stupi-
"fied, and comprehend nothing, nei-

"ther

"ther do they know how to defend
"themselves, or how to form; there
"is nothing but confusion; and if by
"accident any orders are given, which
"is seldom done, you are to speak to
"men who are deaf and immoveable.
"Being little used to military exercises,
"or submission, or to the obedience
"due to their officers, they throw
"away their fire in the air, and are sure
"to be beaten; and that, because the
"soldier is not used to obey, and be-
"cause we are never prompt enough
"in punishing; but especially, be-
"cause young officers neither know
"how to command, nor make them-
"selves obeyed; and those that do, often
"dare not, lest they should incur the
"hatred of their companions, who ima-
"gine that punishment makes soldiers
"desert."

Such was the opinion of one of the greatest Captains France ever had; an opinion founded on experience and compleat knowledge, and which may furnish the best lessons to officers who will reflect thereon. The decay of discipline

was

was at all times the loſs of ſoldiers, and the cauſe of ſhame to officers, who diſhonour themſelves leſs by the defect of courage, than by their neglecting to apply themſelves to their profeſſion.

" You muſt not imagine," ſays this ſame General, " that ſubordination and
" ſervile obedience leſſen a man's cou-
" rage : it has been always ſeen, that
" the ſevereſt diſcipline has been eſta-
" bliſhed, where the greateſt exploits
" have been done *."

Therefore an officer, who marches at the head of a body, ought to keep up the moſt exact order, and a profound ſilence, ſo that they may be always prepared to execute whatever movements he orders for their defence, however dangerous they may be. But in giving theſe orders, he ought always to appear confident and determined, which makes the ſoldiers think he is ſure of his aim, and that he has taken the beſt meaſures.

Soldiers, when they ſee their chief wavering and doubtful in his commands,

* Chap. I. of Diſcipline, *Reveries.*

mánds, imagine him to be at his wit's end; and seeing him disturbed, they themselves will be affected in the same manner.

It is on these occasions, that an officer ought to keep his head clear, to be able to manage his party, and to be obeyed directly. The danger is much greater on a march, than in an attack; in the last, the soldiers have their arms in their hands, and seeing the enemy near, they are always ready to engage; the contrary is observed on a march, they are less on their guard, and have not, in like manner, their arms in their hands. " At such a " time," says Vegetius *, " an attack " stuns them, an ambuscade confounds " them." An officer, who has a mind to put himself out of danger of these surprises, ought, when he gets at a little distance from the camp, to make a corporal or serjeant march eighteen or twenty paces before him, with four or five men, more or less, according to the strength of his detachment: and also

two

* Lib III. Chap. II. of Institutions.

two or three at an equal diftance on his flanks, to make difcoveries, and to examine hollow ways, fwamps, lakes, or ditches, which are on the right and left of the road; to fearch into farm-houfes, barns, mills, and other places where the enemy might lie in ambufh. He ought alfo to ftop all peafants, men or women, who are going the fame road, and endeavour to pafs him; and fhould march them with him, till he is paft all danger. I could mention feveral officers of different ranks, who were furprifed, beat, or carried off, by neglecting fome of thefe precautions; but I will confine myfelf to this example, which fuits well to my purpofe *.

During the war in Spain in 1674, Marfhal Schomberg, who commanded the French army, having a mind to cover Roufillon, ordered a confiderable detachment to march, to fecure the convoys that were coming from Perpignan

to

* The late war in Flanders, and the prefent in America, have furnifhed fome fcandalous, and fome very unhappy inftances, of this want of caution in officers, who fhewed thereby, that their rank was much higher than their judgment

to the village of St. John de Payés, about three leagues from Perpignan.

This corps was posted on a height which was near the high road, from whence the commanding officer sent his lieutenant and thirty men to take possession of a chapel that was on an eminence still higher, at about three hundred paces from his post; from which eminence the lieutenant could easily discover the Spanish incampment on the plain of Boulou, and over which their parties must pass to attempt our convoys.

From Boulou to these two posts there was a long hollow way, through which the enemy might march under cover; and as surprizes were to be feared every day, there was a detachment posted also in a cottage called the *red house*, with orders to light fires to give notice to the other guards, if they made any discoveries, and to be always ready to help one another.

A Spanish officer, with forty horse, passing the hollow way under favour of the night, and being well acquainted

with the country, and the position of the guards, lay in ambush in the middle of the three posts, with a design to surprize the lieutenant's detachment, who went every morning to relieve the guard of the chapel. This lieutenant having got into the hollow way, thro' which he was necessarily to pass, the enemy fell upon him, and charged him so roughly, that all his men were either killed or wounded, before he had time to recollect himself. He received, for his own part, two cuts of a broad sword on the head, from the Spanish officer, who added to this treatment these insulting words: "Go, says he, learn an-
"other time to do your duty better,
"and to reconnoitre a place where you
"are to pass with your guard."

I will not comment on this passage, which is taken from the relation of the Catalonian war; because I believe it will be sufficient to read it once, to prove what I said, that it is necessary to examine every place where the enemy may lie in ambush to surprize you. But as it is difficult, or rather impossible, for a detachment that marches in a suspected country, to examine all the villages by which they must pass, and where the inhabitants are oftentimes as much to be dreaded as the enemy; I think if an officer can avoid them, he had best turn off a little, and come in again to the road when he has passed them.

It is obvious, that for making these discoveries, none should be employed but the oldest soldiers of the party, whom you must order never to stop to drink, to divert themselves or talk with the peasants, and never to lose sight of the detachment; but to stop all persons that endeavour to pass before them, and to come immediately to give an account of what they saw or heard to the commandant.

But as all the precautions that I have mentioned, do not remove the possibility of an officer's being attacked upon his march; he must, as soon as he sees the enemy, examine whether their party be greater than his own; whether it is horse or foot, or both together. If they are

are cavalry, and superior to him in number, he is not to be discouraged on that account; but, on the contrary, he ought to avail himself of his own advantages, by throwing himself into a close country, uneven, or cut, which may be difficult or inaccessible to the horse. He must also raise the spirits of his men by resolute and bold expressions, and endeavour to make himself master of some post, where he may be able to maintain himself, while he sends a faithful soldier to inform the General of his situation. If in this situation the enemy marches towards him, he must do his endeavours to support the efforts of their attack, ordering his men not to be in a hurry, but to save their shot, and not to fire till they can reach the enemy with their bayonets. However contemptible natural fortifications may appear, such are found in every country by chance, which courageous men have defended with extraordinary valour. The last age shews what seven soldiers could do in one of these situations: The Duke of Rohan says, in his

Memoirs, that they ſtop'd, for two whole days, before a poor houſe built of clay near Carlat, a whole army, which Marſhal Themines was leading to the county of Foix, conſiſting of ſeven thouſand foot, and five hundred horſe. If the road, in which a detachment is attacked on its march, be covered on either ſide with vines, wood, rocks, or by ſuch rough broken ground as may prevent the cavalry from penetrating it; an officer, as I ſaid, ought to throw himſelf into it directly, and to continue his march towards his poſt by that way, keeping his men cloſe together, and always ready to receive the enemy.

If, on the other hand, the party of horſe which he perceives, be pretty near equal in ſtrength to his own detachment, he is not to diſcontinue his march on that account; but ſhould form his men into a cloſe platoon of five files of ſix men each, if he has thirty; of ſeven files of eight, if he has about fifty; or of ten files of ten, if he has one hundred; and thus, with their bayonets fixed,

fixed, presenting their arms on every side, he is to continue to march towards his post. An officer, who marches in this manner, without losing his order, and in silence, will convince the enemy, that he is not afraid they should fall upon him. But, however, if they should, he must halt his men, make his first rank kneel, pointing their bayonets to the horses breasts; the second must kneel also, presenting their arms; and the third shall take aim over their heads. You must observe here, that I only speak of a detachment of thirty men; for if the body is greater, you may make two ranks aim at the same time. in this case, an officer must forbid his men to fire till the enemy's horse are within ten paces of the bayonets of the first rank; then the standing rank or ranks which took aim, or were presented, are to give fire, and to reload immediately; those of the second rank are to stand up at the same instant, and present, in order to fire, if the officer commands them; but if the first or second fire has disconcerted the enemy,
he

he muſt order his men to riſe, and continue his march; always ready to begin again, if the enemy ſhould return.

But if the enemy's party diſcovered be ſuperior, conſiſting both of horſe and foot, or of foot only; the officer muſt endeavour to make himſelf maſter of a mill, or a ſingle farm-houſe, to defend himſelf till his General, to whom he has given notice, ſends to diſengage him. If he ſees no way to poſſeſs himſelf of an advantageous poſt, or get to the place he is detached for, he can do nothing better, than to fight his way retreating, and return to the camp, in coaſting along a river, or wood *, if he can,

* The tranſlator thinks the wood is preferable to a river, eſpecially if it be cavalry that oppoſes them, and that they can get that way to their poſt or camp, and would recommend the following method, partly copied from that practiſed by the Indians in North America; which is, as ſoon as they get near the wood to break entirely, and ruſh in with their firelocks in their hands, letting them ſwing at arms length, ſo as to avoid the branches of trees; and when they are got into the thick part ſeven or eight yards, they are to halt: it muſt be obſerved, that they will probably get in at ſeveral diſtances from each other, and ſo much

can, to avoid being furrounded; and if he is fo clofely purfued, that he cannot avoid being beaten and taken; I fee no better expedient to be adopted in this cafe, than that of the *Barbetts* * of the vallies of

much the better, as they can, by that means, form the fooner, *in their manner*, which is in files of five or fix men, or more, according to their ftrength, leaving a fpace between every file, of two feet or more; fo that when they find the enemy near them, and good cover for themfelves, they face about, and form, as the place will allow them, either in one or more ranks, or they advance by fingle files, to fire on the enemy, endeavouring always to cover themfelves behind trees, or ftones, if they can; and they may advance unperceived, by creeping on their bellies, and by this means the enemy will be often deterred from purfuit, not knowing where they may meet refiftance, or what to fire at; and the beft way to deceive them is, when you are fairly in, for the whole to lie down, and creep, as the officer fhall direct; and if the enemy perfift in purfuing you, after making your beft defence, you muft retreat in the order of open files, ftopping every now and then to amufe them with a fire, which will greatly annoy and delay them

* *Barbetts* are peafants fubject to the King of Sardinia, who abandon their dwellings when the enemy has taken poffeffion of them The King forms them into bodies, who defend the Alps, being part of his dominions.

[88]

of Piedmont, who disperse themselves, and retiring from tree to tree, or from rock to rock, so harrass their pursuers, that they can neither beat them, nor take one man.

I promised to mention some remarkably clever *manœuvres* of private officers, which may serve as proofs and explanations of the articles that I treat of. Therefore I cannot pass over a march of M. de Beuvrigny, captain in the regiment of Cambresis, which would do honour to a general officer: I take it from the history of the revolutions of Genoa.

During the Corsican war in 1737 and 1738, the king sent reinforcements to the island, to reduce the malecontents to reason. A convoy, escorted by a frigate and two armed barks, appeared in the beginning of 1739, steering towards San-Fiorenzo; but they met with a dreadful storm the 8th of January, which dispersed them; nevertheless, all the vessels of this convoy arrived with four French battalions, at different ports of the island, except two tartanes, which
had

had the misfortune to run on fhore the fame day, on the coaſt of the province of Balagna, to the left of the river Oſtregone. M. de Beuvrigny, who commanded fix companies of the regiment of Cambreſis, which were embarked in theſe tartanes, ſaved himſelf and the troops, by his preſence of mind and reſolution.

It was ten o'clock at night, when the tartane, that this officer was on board of, ſtruck againſt the rocks, with a dreadful ſhock, about one hundred paces from the coaſt: he hindered his people from leaping into the ſea, where they muſt inevitably have periſhed; and with a piſtol in his hand, he compelled the ſailors to launch their boat, and did not ſave himſelf, till after embarking ſucceſſively all the ſailors and ſoldiers, which took up above two hours.

He had no ſooner got on ſhore with his three companies, than he had intelligence brought him, with advice to think of his ſafety; for if he ſtayed till day-light, he run a riſque of being attacked

tacked by the Corsicans: but he would not abandon the three companies that were in the other tartane, which was also stranded on a sand bank, at a little distance from the first; the boat of which was lost, in carrying some of the officers and soldiers on shore, whose bodies M. de Beuvrigny knew on the strand. He determined to assist those that were still in the vessel, and made his men go into some cottages to warm and rest themselves for the remainder of the night. At day break he sent the boat to land his comrades, who brought on shore about sixty firelocks, and a hundred and sixty charges of ammunition; half the muskets were without locks, being taken off to prevent accidents on board.

M. de Beuvrigny having reviewed his men, who amounted to one hundred and forty only, posted the soldiers, without arms, in the middle; on the flanks, those who had muskets, without locks, but with bayonets fixed; and in the front and rear, those who had pieces in order to fire. After this prudent disposition,

position, he set out for San-Fiorenzo, from which he was about five leagues distant; but he soon had the Corsicans at his heels, who had heard of the ships wrecked on their coast. M. de Beuvrigny crossed the river d'Ostrigone, in their sight, having the water up to his middle, and continued his route by a mountain, in spight of their shot, which he answered now and then. He killed a good many of the Corsicans, and had some of his own men wounded; but in spight of the good care he took of his ammunition, it was soon spent, so that he had but five cartridges remaining among the whole party; and had hardly got half way, when a great body of Corsicans appeared, of horse and foot, preparing to surround him, and to put all to the sword: night approaching, his men overcome with fatigue, without guides, and without ammunition, seeing no other remedy, he determined to surrender himself prisoner. The French General greatly commended the bravery of this officer; who being reclaimed in the King's name, was immediately set at liberty, with all his men.

The conduct of M. de Beuvrigny was so prudent, and so well concerted, that though it was not succesful in the end, I thought it my duty to mention the whole; his presence of mind in the shipwreck, his zeal to save his soldiers, his good dispositions in his retreat to San-Fiorenzo. He withstood the repeated attacks of the Corsicans for a long time, and would certainly have retreated in good order to that place, if his ammunition had not failed, and if he had not had cold, hunger, thirst and fatigue, and a rebel army in a dark night, to contend with, in the midst of a revolted and unknown country.

C H A P. VI.

Of the establishment of a body in a post.

THE moment of the establishing a body in a post, is the most critical a detachment can be in. We have often

often seen officers attacked at the inftant when they thought they had nothing to do, but to take, at their leifure, the proper meafures to remain with fecurity in the poft they had taken poffeffion of.

If the body that arrives at a poft is to relieve another, the officer to be relieved is to put his guard under arms the moment the centinels have informed him that a new detachment is in fight. This detachment being examined, he may let them enter, in order to take poffeffion of the poft, in the room of thofe that are going out. The corporals are to go immediately and relieve the centinels, and the officers and ferjeants are to give each other the neceffary orders to be obferved in the poft, both for day and night: fometimes thofe orders are given in writing, but often are verbal. An officer, who commands in a poft, fhould take the greateft care to remember them. He muft alfo enquire of the officer he relieves, whether the enemy make incurfions into his neighbourhood? Whether their parties are far off? Horfe or foot? And their fituation.

situation? After having procured the exactest information of these things, he may take the best precautions to put his post out of danger of surprize.

As soon as the corporals are returned from relieving the centries, the officer who is going off, is to form his party into ranks, and march back to camp, with the same precautions that he set out. The new guard are to remain under arms until the old one has got off about twenty or thirty paces: then if they are posted in a redoubt, the commander must make his men lay their arms on the top of the parapet, and see that they cover the locks with their haver-sacks, in order to preserve both them and the priming from dust, dirt, or moisture.*

But if an officer has relieved a party in the open field, in a post without any fortifications, and where he does not chuse to make any, he must order his men

---

* However it seems better, instead of laying the muskets, as our author directs, on the top of the parapet, to rest them on the *banquette*, leaning their barrels against the parapet.

men to ground their arms in the day time, and not to ſtir far from them; and to keep them between their knees at night, as they ſit round their fire,* obſerving to turn the locks inwards, for fear of accidents. Theſe precautions being taken, the officer is to go viſit the centries, and the places round his poſt, to know where to retreat to, in caſe of an attack.

But if an officer is detached to a poſt that has not been occupied before, as ſoon as he arrives, he muſt poſt his centries to prevent his being ſurprized from without,

---

* The tranſlator thinks theſe fires very dangerous for a party in an open country, without a redoubt; unleſs you poſt centries at a conſiderable diſtance from the fire, round about you, to obſerve the approach of an enemy; for thoſe at the fire can ſee nothing at a diſtance, ſo that the enemy may get between them and their camp, or place of retreat, unperceived, even till they are within twelve or fourteen yards of the fire; nor can they be heard, as the burning of the wood and wind together, make a loud noiſe; and when got ſo cloſe, a very ſmall party will (by boldly running in) with bayonets fixed, carry off a much larger party ſitting down, and who have not half a minute to riſe, and too confuſed to think.

without, and place his arms as I directed before. But if he is sent to a mill, a house, or cottage, he must draw up his men in order of battle about fifteen or twenty paces from the place, and send a serjeant or corporal, with five or six men, to examine the cellars, the chambers, and garrets; which being done, he may post his centries, and take possession of the place; and make every soldier place his arms so, that each may find his own without confusion: He must lodge the inhabitants elsewhere, and intrench himself in the manner before-mentioned for single houses.

Lastly, if an officer is to establish himself in a village, as it would be very difficult for him to examine every house and place where the enemy might lie concealed, he should send to the magistrate and people of the first rank, keeping his men drawn up under arms, at about fifteen paces from the village, and oblige them to declare, in the King's name, whether there are any hostile parties, suspected people, or concealed arms, in the place; after
which

which, he is to send to post his centries, and enter the village, and post small detachments at the avenues of five or six men, according to his strength; then examine the *chateau* and church, or any other building that stands by itself, for fixing his chief post, and the place of his last retreat, if he should be forced in his advanced posts.

After an officer has taken possession of a post, he should go and see how the corporals have posted the centries; if he finds they are improperly placed, he is to change them. When many centries are to be posted, care should be taken that the oldest soldiers, or those that know their duty best, should be posted at the most exposed and distant places, and in such a manner, that they may discover all approaches to the post; sometimes he may place a few in trees, so that they may see a far off, and not be seen themselves by the enemy.

After the officer has made this round, he is to visit the places about his post, to see if it be necessary to cut up any roads,

roads, or to stop them by *abbatis*, to pitch on the places where trenches may be dug, and what sort of intrenchment he ought to make to secure himself, and how he may avail himself of all the little contrivances before-mentioned. If there should be near at hand a hollow way, or a thicket of wood, a house, or any other covered place, which the enemy may take possession of, and where they may lie in ambush, in order to fall afterwards on the post; here he should place a small guard of six or seven men, commanded by a serjeant or corporal, who are to have orders to fall back to the chief post if they should be attacked, or to support themselves till they can receive assistance. The men of this little guard must be charged to make no fires, because the light, or smoke, would be a guide to the enemy, who would chuse to avoid it, if they intended to surprise the principal post. Experienced officers and soldiers light fires in places where they have no guards, to make the enemy imagine they have posts every where, and make

their

their ambuscades in places where they have no fires. This *finesse* may also be used in all posts that lie in an open country, from which the officer, to this end, must detach two or three men to go, during the night, from one fire to another to keep them up.

After this external arrangement, and when the centries are posted at the avenues, on the bridges, and the tops of the steeples, the officer is to see what sort of intrenchment is best for the defence of the post; and as soon as he is determined in this point, he is to mark them out, and make the workmen finish them; then he is to place therein the little guards, which are appointed for their defence; and if the detachment is to continue for many days, he is to lodge the men in the nearest houses, giving order to the magistrate to furnish them with straw. Care must also be taken, that the soldiers so lodged be not suffered to stray from the houses, that they always lay their arms where they can find them easily, and, if possible, there ought to be an officer to command at every one of those posts, or a serjeant or corporal at least; the soldiers who guard the intrenchment or fort, are to sit all night on the *banquette*, leaning their muskets on or against the parapet, and to take care to be alert all the time. Let the enemy be at what distance they will, the officer ought never to sleep but with his cloaths on, so that he may be always ready to go wherever his presence may be wanting; and to make his serjeants and corporals go the rounds often to visit the centries *.

If he has chosen the *chateau* or *mansion-house*, the church, or parsonage, or any other house, for the chief post of the village, and that these places be inhabited, the inhabitants must be lodged elsewhere, so that no body should stay in the post to hinder or betray him, or any

* As nothing can be more injurious to the character of an officer than being surprized in his post, and as serjeants or corporals are not always to be depended on, the translator would recommend it to the officer to go the rounds himself two or three times a night, without mentioning his hours; and to order the serjeants or corporals to go at least every hour.

any way obstruct him in carrying on whatever works he may think necessary. An officer should never suffer himself to think, that the little time a detachment is to stay at a post, should be a reason to neglect preparations for its defence. " When an officer is in a " post," says M. Vauban *, " he should " intrench himself immediately, tho' " he were to stay but four hours."

To which I will add, that all his works ought to be well made, and so contrived, as to defend every side where the enemy may come.

Monsieur Follard gives us on this subject an excellent maxim, which may serve as a general rule. " You must," says this author †, " *imagine* that your " post is attacked, in order to *imagine* " the method of defending it." And M. le Baron de Travers, in his Observations on the Art of War by Monsf. de Puffegur, says ‡, " that posts ought
" always

---

* Attack and Defence of Places, Tom. II. page 180.
† Tom. V. Defence of Posts.
‡ Chap. 10. second part.

" always to have strength and means
" of defence, in proportion to what the
" enemy may employ against them."

M. Duclaux de Barrieres, captain in the regiment of Lorrain, beyond any one I ever saw, seemed persuaded of the utility of speedily intrenching his party; when this officer was to stay a few hours in a post, he immediately made an *abbatis* of trees; and if he was in a village, he directly intrenched the *chateau* or principal house.

C H A P. VII.

Of precautions to be taken in a post, to avoid a surprize.

I HAVE said, that the security of an army depends on the vigilance of its advanced guards: And is it not a matter of the highest importance to a General, supposing him to have made the best disposition possible, that the officers detached from his army should know

know how to go about, and be sure to execute, whatever was committed to their charge? " The chief object that " a soldier ought to have in view, when " he is detached," says Monsieur de Vauban, in his Treatise of War, " is " always to foresee any bad accident " that may happen to him." The neglect of regularity, or the least relaxation in duty on a post, may have very unhappy consequences. History furnishes us with many examples, where camps have been surprized, and armies cut in pieces, by the neglect of detachments, who ought to have watched for their preservation.

After an officer is established in his post, his principal concern should be, how to provide against an unexpected attack, his being betrayed, or carried off.

M. de Travers says *, " the only " means in war, to be safe from sur- " prizes, consist in taking precautions " against every thing the enemy can " possibly undertake; therefore you " are

* Supplement to Military Study.

" are not to reckon that your safety
" depends on the probable, but on the
" possible distance of the enemy."

To avoid the ill consequences that may result from neglect, an officer, when he is sent to a redoubt, or any other detached post, should let no strangers whatever enter into it, not even any soldier that did not belong to his detachment: he should forbid all his own people to pass the bounds * where the centries are fixed, or to straggle from the post, under any pretence whatsoever; and he should call over the roll three or four times a day.

When the corporal is going to relieve the centries, the officer is to examine the soldiers who are to relieve them, before they march off: he is to visit them after they are posted, and to make his serjeants and corporals visit them now and then.

When it grows dark, he is to make the centries come nearer to his post, so that by forming a less circle, nothing

may

---

* By order of the 1st of July 1727, every soldiers, who passes the limits, shall be hanged.

may pass unperceived between two of them. If there are signals to be made, or to answer, they must be ordered to be very attentive. If, in the neighbourhood of the post, some parts are much wooded, in such places two centries ought to be posted, with strict orders not to speak to each other, or walk about; and some of them may, as I said before, be posted in trees, minding to relieve them every two hours, or every hour, if the weather is severe. If an officer, in making the round of his centinels, finds any new soldiers among them, he should remind them of the duty they are bound to observe while they stand centry, and tell them they ought never to quit their post, nor fall asleep upon it; and that they are not to suffer themselves to be relieved by any but their own corporals; and to permit no soldier to go from their post, and to inform their officers of any thing they see; to stop any body that may advance towards them, till they know them, and to fire on those who refuse to answer, after calling out three times, *who comes there?* laftly, if they find them ftill approaching, they are to return to the poft of their commanding officer.

It is not fufficient for an officer, who commands in a poft, merely to receive reports from his centinels; but he fhould, as it were, for paftime, fhew his detachment the way to defend themfelves, if they fhould be attacked; and explain to them, if the enemy made fuch an attempt, how he would oppofe them by another; if they undertook this, he would ufe that, and baffle them in every point. He may make fome of his men try to fcale the works, to fhew all his people the difficulty of executing it. By exercifing them in this manner, he will prepare them to refift the enemy with eafe: thus he will give them a high opinion of themfelves, and make them put great confidence in him. He muft avoid all this time, while he treats them as comrades, not to be too familiar with them; for in that cafe, if in the midft of an attack he fhould order them to do fomething that was not to their liking, inftead of obedience, they might

might perhaps mutiny, and disobey his orders.

After the taking of Bellegarde in Roussillon, by Marshal Schomberg, in 1675, there happened an event, which shews the importance of what I now say: I take it from the relation of the wars of Catalonia; I will give the whole, being analogous to my subject.

At a league from this fortress, on the road to Colioure, there was a chapel dedicated to the Virgin: this chapel was built in the middle of many rocks, whose points, almost inaccessible, served it for a wall, so that it was very strong, as well by situation, as from the rocks being proof against cannon-shot.

Marshal Schomberg, being resolved to make himself master of this important post, which was defended by a Spanish captain and fifty Germans, detached a large body from the army, commanded by M. de Gassion, marshal de camp: the trenches were opened, some cannon were drawn thither by men, and a battery was erected on a rock near the post; but it had little effect,

effect, though it was well ſerved. The captain and his ſoldiers laughed at the beſiegers for five days, and ſeemed likely to be in the ſame condition for a long time to come; but a cannon-ſhot having killed three Germans, who were looking over the wall, all the reſt loſt their courage, and told their captain, in an inſolent manner, to ſubmit, and make his compoſition, or they would do it for him; he was aſtoniſhed at their cowardice, remonſtrated to them their duty, but in vain; they hung out the white flag. The French, delighted with their diſobedience, conſented to treat with them; but while the Germans were all in confuſion, the French got poſſeſſion of the chapel-gate, before any conditions could be drawn up, and they were all made priſoners.

This example, which ſhews how difficult it is to force a brave man in his poſt; and how important it is for a commander to maintain an awful authority over his own ſoldiers, ſhews alſo that one ought never to deſpair, on account of the revolt of a few mutineers. The spirit of rebellion never seizes a whole body at the same instant; but it is insinuated, and spread by the seditious propositions * of a few. As soon as an officer perceives this, he should command immediate silence; and if they have the insolence to continue the tumult, he should take instantly a musket, shoot the most daring of the mutineers, and threaten to hang all those that shall refuse obedience. I could account for the reasonableness of this conduct, and relate many examples to prove, that this way, though it may appear violent, is the only one to check a mutiny in soldiers, or even in a mob. But as that would be foreign to the work I am upon, I return to my subject.

When an officer has shewn his men the advantage that an intrenched body has over those that attack them openly, he must take care to keep up good order, and not to suffer himself to be ensnared by the enemy.

If

---

* By order, of the 1st of July, 1727, mutinous soldiers are to be sent to the provost to be hanged and strangled.

If an officer is detached towards the limits, and that deserters come to his post, he must be cautious not to let them in. He ought, on the contrary, to make them remain without on the glacis, and send two or three soldiers to receive their arms, and conduct them to the General of the army, escorted by some musketeers. If these deserters are very numerous, and that he cannot spare men enough to conduct them safely, without weakening his own post too much, he must write to the Adjutant-General, to beg of him to send a detachment to receive them.*

It

* If those deserters should offer themselves at an advanced post, near or on the glacis, in the time of a siege, and are too strong to be received in the post; if the above method is followed to send men to receive their arms, or to keep them there till the General can send to receive them, the whole will, in all probability, be knocked in the head from the couvert way, or be retaken. Therefore, if they were ordered, on their approach to the post, to throw down their arms; when that's done, the officer may send as many armed men, as he can spare, to conduct them to the trenches.

It is not alone sufficient to take the steps that I have shewn for the internal security of a post that is to be defended. But a skilful officer ought also to look to the external safety, and endeavour to discover the designs that the enemy may form against him. The most critical time for detached officers to be alert, is an hour or two before day; great care must be taken to keep the soldiers awake, and to make them sit on the *banquette*, each close to his musket. One or two patroles ought to be sent out, during the night, and at day break, as scouts, to make discoveries in the environs. Those patroles should be of four or five men, and be ordered to march slowly, and with the least noise possible, to examine the hollow ways, hedges, ditches, woods, and neighbouring houses, to stop and hearken every now and then, to hear if there be any noise, and to return in half an hour, so that another patrole may go out immediately after them.

It sometimes happens, when two armies are encamped opposite each other,

and

and that there are many posts in one and the same line, that two night patroles meet each other. Then, as it is not possible to discern whether they are friends or foes, that patrole, which has first seen the other, should slip on one side of the road and lie down behind some bushes, or in a ditch, so as to observe their approach, and examine if the other be the strongest: if so, they are to let them pass, without saying a word, and then to return to their post, to give intelligence of what they saw. But if, on the contrary, it should be weaker, the officer should make the signal that was given in orders, or that which his officer gave him for the patroles of the night. This signal is commonly to strike one or two strokes on their pouch, or on the stock of their musket; to which the other is to answer by an agreed number according to order. If the other patrole will not answer, the first is to march up to them, with bayonets fixed, and fire on them, if they endeavour to retire, and oblige them to lay down their arms. I saw, during the war

war in Italy in 1745, old foldiers, who, of their own accord, afked leave to go out fcouting, and took a great deal of pleafure in it.

When a detachment is to be pofted oppofite the enemy, one may expect to be attacked; therefore they fhould advance fome little parties during the night, about twenty-five or thirty paces from the poft, who are to lie down on their bellies, between the centries, at the places where they think the enemy may come: and orders fhould be given to the commanders of thofe little bodies, to make a fingle foldier reconnoitre any parties they may fee; fo that they may not miftake the friendly patroles for parties of the enemy; and all fhould return to their chief poft on the firft report of a mufket.

When one is charged with the defence of a poft that ftands fingle, it is of the greateft confequence not to neglect any of the precautions that I have before mentioned; but there are others as effential to be known, when a village or hamlet is to be defended. An officer

officer sent to a post of this kind, should be careful to prevent any suspected persons from slipping in, and to keep the peasants from revolting. To this end he should order, by means of the magistrate, two of the most noted peasants in the place, to be posted with the centries on duty, at the two only entrances into the place, all others being stopp'd up by the entrenchment. Those peasants are to be relieved every two hours, and are to be ordered to examine the inhabitants that may come in or go out of the village; and the whole of the inhabitants shall be declared responsible for any accidents that may happen thro' the treachery of these peasant centinels, or if thro' their neglect any of the enemy should get into the village disguised.

It should be given in orders to the soldiers who guard the intrenchments, not to suffer any peasants to approach them; and to those who are posted at the passes, to stop them up at night, by laying two trees across, and not to open them till next day; and also to search all waggons loaded with hay or straw,

casks,

casks, or any thing else, by thrusting in iron spits, or their swords, to feel that there are no men, arms, or ammunition concealed therein.

During all the time that a detachment occupies a village, the inhabitants must not be allowed to hold fairs or markets, or make processions; for often those assemblies afford opportunity to the enemy, to slip in and surprize the place. Polibius gives us a lesson on this subject, which may not be found disagreeable here.

" The unhappy consequences of this
" liberty have been experienced above a
" hundred times," says the * translator
of this author, " and yet no remedy is
" applied. How unjustly does man
" pass for the most artful of all animals?
" There are none more easily surprized.
" For how many camps, how many
" garrisons, and how many posts have
" been lost by this liberty? This mis-
" fortune has happened to an infinite
" number, and nevertheless we are still
" novices, as to this sort of surprize."

*Don Vincent Thuillier, t. 5. liv. 16.

An officer, who commands in a post cannot be too watchful, lest there should be any plots against his safety. The enterprize of the enemies on Brissac, in the month of November, in 1704, comes too nigh my subject, to be passed over in silence. The Governor of Fribourg, having formed a design against that place, set out at night on the 9th or 10th of that month, with two thousand men, a great many waggons; some of which were loaded with arms, grenades, firelocks and pitch, and the others with chosen soldiers. All these waggons were conducted by officers, disguised like waggoners, and were covered with poles, with hay over them; so that they seemed to be only loaded with contribution hay. In this manner, and being favoured by a thick fog, they arrived at eight o'clock in the morning, at the new gate. Two of the waggons with men, and one with arms, got into the town immediately: but an Irishman, overseer for the undertakers of the fortifications, seeing thirty men near the gate, who had not the appearance

ance of peasants, though they had the dress, asked them who they were? and why they did not go to work, like the others? Upon their not answering, or appearing confounded, he struck some of them with his cane; upon which, the disguised officers seized the arms that were in the waggon next them, and fired fifteen or twenty shots at him, within five or six paces distance, without wounding him; he threw himself into the ditch immediately, where they fired several shot more at him, to no purpose, while he cried out loudly, *to arms*. At this noise, the advanced guard of the half moon, and that of the gate, took arms, and tried to hale up the bridges; but they could not, on account of the waggons that the enemy had stopped on them. The officers and soldiers, who were in the waggons, leaped down, and took to arms; and joining the rest, attacked the guard commanded by a Captain of grenadiers, but they were repulsed; and five being killed, the rest were disheartened, and fled, some into the town, and some out: then he shut the

the first gate, through which, being of pallisadoes, or rails, the enemy, who were on the bridge, fired at every thing that appeared. The Captain left there one half of his guard, and mounting the ramparts with the rest, kept a continual fire on the enemy. A Lieutenant, who commanded twelve men in an advanced guard, was attacked at the same time, by an officer who clapp'd a pistol to his breast; but he wrenched it out of his hand, and shot him dead. This Lieutenant defended himself till the end of the action; but having received many wounds, he died the same day. On the noise of this surprize, M. de Raouffet, commandant of the place, distributed the garrison in all the necessary posts, and did every thing that might be expected from a good and experienced officer. In fine, the enemy seeing their design fail, retired in disorder, leaving behind them a great many waggons, and above forty soldiers, who were either killed or wounded. Such was the enterprize on Brissac, which only failed by an effect of meer chance.

The

The excellent *manœuvre* of M. Vedel captain, and since lieutenant-colonel, of the regiment of the Isle of France, is a more recent example, to shew of what importance it is to an officer, to use precautions when he is detached to a post. During the troubles in the island of Corsica in 1739, this officer being detached to Chisoni, a village in that island, the parson of the parish asked the officer commanding in the open country, to permit the penitents of a neighbouring convent to come to this village, according, as he said, to their annual custom, of making a procession to a certain chapel in the place, which he named. The commandant consented; but M. Vedel, who was detached thither with fifty men of his regiment, being surprised to see so numerous a procession coming into the village, which was composed only of the peasants of a revolted country, called *to arms*, drew up his men, and thus disconcerted their projects; in fact, many of these penitents, whom they seized, were found to be armed with
pistols

pistols and swords; an account whereof being sent to Marshal Maillebois, then General of the French troops in Corsica, he commended the activity of M. Vedel, and ordered some of the penitents, parson and all, to be hanged up on the spot.

This example, and a thousand others that I could cite, shew, that an officer who commands in a post, ought to be very attentive, lest he should fall into the snares the enemy may lay for him: the loss of a post, though it may appear very trifling, may have the worst of consequences. The surprise of Amiens in Picardy, in 1597, will ever be remembered. The Spaniards having formed a design to surprise this town, laid some soldiers, disguised like peasants, in ambush, in a house near the gate, and sent forward a waggon loaded with walnuts, of which the driver spilt a sack, as if by accident, and whilst the soldiers of the guard ran crowding to gather them up, the disguised Spaniards fell on them sword in hand, surprised the gate, and made themselves masters

of

of the town, which coſt Henry IV. ſix months and a half ſiege. Theſe events, which are frequent in hiſtory, ſhew how alert an officer ought always to be; the many ways there are to ſurpriſe a poſt, ought to make one more miſtruſtful, to be always ready to parry any blow, and to make it abortive.

If any ſtrangers or neighbouring peaſants come to ſee their friends or relations, the centries ſhould ſtop them, and ſend notice to the commandant, who is not to let them in, till after the chief man of the place, the parſon, the magiſtrate, or ſome of the moſt conſiderable inhabitants, promiſe to be anſwerable for them. Note alſo, that this permiſſion ſhould only be granted on working days, not on Sundays or holidays, as all the peaſants are idle on thoſe days. This precaution, relating to the enemy without, is eſſential; but another, which is no leſs ſo, is to be on one's guard againſt the inhabitants of the village itſelf, who, in an enemy's country, are always ready to betray, or to revolt. The commandant of the detachment ought

ought to take one or two of the magistrate's children, or three or four from the moſt noted families, whom he is to keep in his principal poſt, as pledges of the fidelity of the inhabitants: Great care ſhould be taken that theſe children ſhould receive no ill treatment; that they ſhould not be detained longer than half a day each; and that they be relieved by others. The commandant ſhould forbid the inhabitants to gather together in public-houſes, or the walks, or any where elſe; and ſhould have this order paſted up over the church door. If, after they come out of church, they ſhould ſtop in the open places to talk together, he ſhould ſend the patroles to make them diſperſe. Orders ſhould be ſent to all the inn-keepers, and all the inhabitants, not to receive any ſtranger into their houſes, without ſending notice thereof to the commandant: and they muſt be ordered not to ſtir out of doors after the drum has beaten the *retreat*, on pain of being ſhot by the centries who may ſee them; or ſeized and carried to the black-hole by the

the patroles, who should march slowly, and stop now and then to hearken if there be any noise; to go through all the quarters that shall be appointed for them; and go to give an account to the commandant of what they discover, that may occasion any alarm in the post.

If a quarter of the town should be on fire, or if the inhabitants should quarrel among themselves, an officer should then be very cautious how he sends his guard to assist them, as those are often snares of the enemy to try to divide the forces of a detachment, in order to attack them afterwards. A commandant, on the contrary, should order the alarm-bell to be rung, and make all the posts, that defend the village, to take up arms; and order those who command them, to make their men stand to the parapet with their arms, to watch all that passes without side the village. The soldiers of the chief post should also be under arms, and the commandant should, at the same time, send four or five men with a serjeant or corporal, to part the fray, or make the inhabitants work to put out the fire.

As all the precautions neceffary for the fafety of a poft are too many to be remembered, an officer fhould give his orders in writing, and have them pafted up, particularly at every little feparate poft or guard.

Officers, on detachment in villages, fhould efpecially be careful not to opprefs the inhabitants, by demanding exorbitant fupplies. I know that it is fometimes permitted, by an order of the * General, to exact fire-wood, forage, candles and oil, for the feveral guards; but thefe demands ought to be proportioned to the abilities of the inhabitants. I could mention here many examples of officers, who have, in a bafe manner, abufed this power, by increafing, to fuch a degree, thofe contributions, that the magiftrates have been obliged to pay them in money, not being able to
fupply

---

* By an order of the 30th of November 1710, it was forbid to exact any thing in the villages, without paying for it.

supply them in kind. An officer therefore cannot be too delicate on these occasions; and should see that the inhabitants are not pillaged * or ill-treated by the soldiers. Every thing is to be dreaded from enraged people; and if the loss of our wealth makes us loose our senses, as is said, to what despair will not people be drove, who seeing their country ravaged, their effects pillaged, and lastly their persons abused, and treated like slaves? I will not say that humanity calls aloud against such rigorous treatment, because it is too common to see war silence the laws of humanity; but I will say, that not only small detachments, but even whole garrisons, have been driven out, and had their throats cut, in the towns they defended, by the inhabitants, whom they had reduced to despair.

History abounds with examples of this kind; but without enumerating them here, I will confine myself to that
which

---

* By an order of the 8th of April 1718, it was forbid, on pain of death, to soldiers, to steal any thing, to scale walls, or break open houses, &c.

which the town of Genoa exhibited to all Europe, at the end of the year 1745.

The Auſtrians having made themſelves maſters of this capital, the Marquis de Botta was appointed commandant, and had under his orders a large garriſon of Germans, who treated the Genoeſe with all the rigour imaginable; while, by orders from the court, they loaded them with contributions. This General having reſolved to take away ſome artillery that was on the ramparts, on the fifth of December in the ſame year, the bed of a mortar broke in a narrow ſtreet; the populace gathered about, but the officer, who inſpected the removal, having ſtruck a Genoeſe with his cane, who ſtood in the way, or refuſed to help, the latter ſtabbed him inſtantly in the belly with his knife. The commotion becoming general, the inhabitants flew to the arſenal, broke open the gates, and took out the arms, repulſed the Auſtrians from ſtreet to ſtreet, and drove them out of their town, after killing above five thouſand of them. A good leſſon, on which I intreat all military people to reflect.

CHAP. VIII.

Of diſpoſitions neceſſary to maintain a party in a poſt.

BUT it is not ſufficient, for the preſervation of a poſt, to have made good intrenchments, and to have taken precautions againſt all kinds of ſurpriſes; for as the enemy may attack it with ſuperior forces, thoſe who are attacked ſhould make their diſpoſitions ſo as not to embarraſs one another, and that every arm may be ſo properly placed, that all may contribute to the common defence.

If it is a redoubt that is to be defended, or whatever other intrenchment of earth, ſeven or eight trees, with all their branches, ſhould be reſerved, to ſtop the breaches that the enemy may make; the parapet ſhould be lined with all the ſoldiers of the party, and arm

arm the first and second ranks with their muskets, and their bayonets fixed, who are not to fire until the enemy are on the glacis; if possible, the third rank should be armed with long weapons, such as spontoons, halberts, lances, forks, or, as M. Follard says in his notes * on Polybius, with long poles, having bayonets fixed at their ends. These long weapons will stop the enemy at the edge of the ditch, or at the outer edge of the parapet, where it will be easy to shoot them. These soldiers, of the third rank, may also be furnished with grenades, or faggots well lighted, to throw among the enemy that have leaped into the ditch; also ashes or slack lime may be thrown on them, the burning dust of which will infallibly blind them. Tho' this last expedient may seem extraordinary, I believe, after many trials, I may take upon me to answer for its success.

It is evident, that the different methods I have been speaking of, to arm soldiers for the defence of a parapet,

cannot

* Tom. III. page 278.

cannot be practised by a small body of thirty or fifty men; this number not being sufficient to form two or three deep, they are to be armed with their muskets and bayonets only; and if the enemy gains the parapet, they must be resisted sword in hand, keeping always quite close to the parapet: care must be also taken to post eight or ten soldiers, more or less, according to your numbers, in the ditch, at the parts the least exposed, and least in sight of the enemy; to keep in this position till the enemy leap into the ditch; then with orders to divide into two parts, one to the right, and the other to the left, to fall on their flanks, with their bayonets fixed. This kind of sally will astonish the enemy greatly, as those who attack, don't dream of being attacked; but, on the contrary, are surprized to see themselves so hotly charged.

The parapet of a redan, is to be lined in the same way as that of a redoubt, observing if the right or left of those redans were joined to any heights, or commanded by any rocks, which often happens; they should be taken possession of by seven or eight soldiers, covered with an *abatis*, to hinder the enemy from making themselves masters of them; and that they should not overwhelm those in the intrenchments, by throwing down heaps of stones.

If it be a *chateau*, a house, a cottage, or a mill fortified with a curving parapet, that is to be defended; a part of the soldiers designed for the defence of the intrenchment, are to be posted, as I have just now said. This first disposition being made, it is not immediately necessary to place soldiers in the ground floor at the loop holes; as they will be useless there while the outwork can be maintained: but if those who defend it, are forced and obliged to abandon it, they are to take refuge in the house, and post themselves at the loop holes. Two of the strongest soldiers are, at the same time, to be placed at each jaum of the door, withinside, with their bayonets fixed, to stab the enemy, who shall attempt to enter, and to pass the defile

made

made by the tree placed before the door.

An officer, who as I said in the Chap. *Of precautions to be taken in a post*, will acquaint his men before-hand, with the different manœuvres that are to be performed, in case they should be attacked, need not be afraid that they should execute, what I am now speaking of, in disorder. The soldiers at the loop holes should never fire, until they are sure of their mark, and mind that one of them keeps the muzzle of his piece always in the loop hole, while the other is charging.

There should be also at the loop holes, of the first story, two or three men to annoy the enemy, by musket shot; and there should be a forked pole left beside each loop-hole, of ten or twelve feet long, to thrust occasionally thro' the holes, to grapple and overturn any ladders the enemy might lean against the walls, observing to push them quick and strong, so as to overturn, at the same time, both the ladders and the men who are on them.

If the windows of the firſt ſtory are not quite ſtopped up, and though the floor is cut away before it, two ſoldiers may be poſted near it, to overturn the enemy's ladders: Laſtly, ſome ſoldiers ſhould be ſent up to the ſecond ſtory, which is generally the uppermoſt in peaſants houſes in the country; they are to be poſted at the brink, the walls where the tiles were taken off, with orders to ſhower down ſtones, aſhes, lime, or half burnt dung, on the beſiegers, and to beat down their ladders with the rafters of the roof, in order to prevent them from gaining the top of the houſe.

If it is a village that is to be defended, and that little guards are poſted at the entrance of the ſtreets, it will be proper to ſhew each of them, in what manner they are to retreat; if being forced, they are obliged to fall back to the principal poſt, defending themſelves from houſe to houſe, and from ſtreet to ſtreet, behind the trenches, that they have cut acroſs them.

If there be a few cavalry in the detachment, they ſhould be poſted in the market

market place, or any open ſtreet, where they may be ready to fall on the enemy ſword in hand, as ſoon as they appear expoſed; but if they are numerous, they may do the ſervice of infantry with ſucceſs.

Laſtly, If there are cannon, they ſhould be placed oppoſite to the ſtreets that lead to the chief poſt, to keep the enemy at a diſtance.

When all theſe diſpoſitions are made, an officer ought to order each and every one of his ſoldiers, to remain at the poſt aſſigned to them; to make a little fire if the ſeaſon is cold, and to place their arms ſo, that they may find them readily, and without confuſion.

CHAP.

CHAP. IX.

The defence of posts.

THE obstinate defence of a post is an action wherein an officer may acquire the greatest glory: this resistance is not so much owing to the number of soldiers destined for its defence, as to the ability of the officer who commands; it is in him chiefly that the strength of the intrenchment exists; and if to his determined bravery the talents necessary on those occasions are also added, and that he knows how to persuade his soldiers, that the treatment they are to expect from the enemy is a thousand times worse than death, one may say, that he will, in some sort, render his post impregnable.

If an officer, posted in a redoubt, is attacked by the enemy, it is not his business to be firing himself; on the contrary, his constant occupation should be to see that his men do their duty well, and that they do not throw away their

their shot idly. If he perceive their ardour cooling in the midst of the attack, he should animate them by his voice; and if he sees the enemy make a greater progress on one side than on the other, let him weaken one to strengthen the other. I know this movement is sometimes dangerous, and that it would be better to have a small reserve to make use of, as occasion may require: but can an officer, who has but a little detachment, scarce sufficient to line the parapet two deep, can he take away twelve or fifteen men to make a reserve?

If the enemy succeeds so far as to make a breach and gain the parapet, two or three trees must be immediately thrown into the breach, with all their branches*, and he must receive the enemy with fixed bayonets.

One

---

* Provided the trees are to be found.—The true meaning is, that you are to stop the breach with trees or stones, or whatever you can, and make the entrance as difficult as possible; and on this account, it is not amiss to provide every field fort with some materials for this purpose.

One may also, as I mentioned in the foregoing chapter, throw handfuls of lime or ashes in their eyes, which will soon oblige them to return to the ditch. Lastly, if the true precaution has been taken of furnishing the hindermost rank with long arms†, and if eight or ten men have been posted in the ditch, so as to come round the redoubt, on right and left, and take the enemy in flank, one need not fear that they will easily master the place, or that their enterprize will cost them little.

But if it be the passage of a river, or a ford, that is to be defended, after throwing, as I said above, several trees with all their branches on the bank, he must there wait resolutely for the enemy, and keep up a smart fire. If they attempt to come down the stream in boats, a good many grenades must be thrown among them; one may also fire upon them with large bird shot, because such shot scattering a good deal, and wounding some in the eyes, some in the face or belly, troubled with so many wounds,

which,

† Such as pikes.

which, though small, are nevertheless painful; the soldiers who have received them, will not fail to raise such confusion in their party, as may make their project miscarry.

If, in the defence of a post, which has some natural strength, and which has been fortified according to the rules I have given, the party should be forced in the first intrenchment, the soldiers should retire into the ground floor, and range themselves at the loop holes; in that instant two men should seize, as I said, on the two jaums of the door, to stop the enemy with their bayonets.

But if it should happen, that the soldiers, placed in the ground floor, should be driven from thence also, one ought not think that the enemy was therefore become masters of the post; these men, forced below, should go up to the next story with ladders, supposing the stairs were broken down; they must draw up the ladders after them, and place themselves at the holes, which should have been made in the floor. If this story should be low enough to reach the enemy through it with the bayonets, a single man will be sufficient to each opening; otherwise there must be two, who must not fire till they can almost touch the enemy with their pieces; these must also be ordered to pour down great tubs full of water, which must have been previously provided there, in order to spill down through the holes in the floor on the enemy that are masters below. This trick, though it may appear singular, is one of the most disagreeable that can be opposed to the assailants; for at the same time that it wets their arms, powder, and their cloaths, it hinders them to see what is passing over their heads, and frustrates any attempts they might make to set the house on fire. If, however, they should penetrate into a room, they must not be suffered to form, or reinforce themselves there, but they must be fallen upon sword in hand, or with fixed bayonets, and, by dint of bad usage, make them renounce the attack. The example I am going to relate proves, that the enemy is always

obliged

obliged to come to this pass when they have to do with brave men.

During the war in Italy in 1705, M. the Chevalier Follard being to defend a little house or cottage, called Bouline, near Brescia, where he commanded four companies of grenadiers: this officer was attacked by all the chosen troops of Prince Eugene's army, who, after firing several cannon shot, and penetrating into the court of the house, were forced to retire. "M. le Prince de Wirtemberg," says this author[*], "who
"feared that we should be succoured,
"thought, that in making himself
"master of a dove-house, from whence
"there was a hot fire kept up, the rest
"of the cottage would not hold out,
"upon which he attacked it; and as
"our soldiers had carried off the door
"to make fire of, the officer who de-
"fended the lower part being wounded,
"could not withstand the fire that they
"made into this door, so was made
"prisoner; there were seven grena-
"diers in the upper part of the dove-
"house,

[*] Commentaries on Polybius, tom. 5.

"house, who were also summoned to
"surrender; but these imagining them-
"selves too well posted to be soon re-
"duced to that necessity, they an-
"swered boldly, that they would not
"give up till the pears were ripe, and
"ready to fall; and that they were very
"capable of proving it. In effect, they
"continued to gall the enemy with
"their musquetry, and did not quit
"their dove-house till the Prince de
"Wirtemberg retired, leaving the place
"covered with his dead."

This defence, which does great honour to M. Follard, and to his brave captains who seconded him, is a good lesson for young officers. M. le Chevalier de Clairac * gives another example, which is no less instructive. In 1742, this officer, marching in the high Palatinate of Bavaria, with a certain number of people, perceived that he was pursued by a troop of Hussars and Pandours, who might attack him with advantage; having examined the different

* See his Treatise of light Fortifications, Chap. III.

ferent avenues of the village of Vurz, where he then was, he barricaded them with waggons, having taken off one or two wheels from each, and with trunks of trees, ladders, &c. he also raised a *banquette* along the walls of the church-yard, where he made a stand with his baggage and his followers, considering the church as a citadel, having broke loop-holes through the door, and the steeple as an enclosure for his last retreat; but there were two houses that almost touched the church-yard wall, and as they were built on lower ground, the tops of their walls were no higher than that, which served him as a parapet; nevertheless he did not chuse to open these houses; but as he was obliged to have a communication with them, in order to avoid being fired upon from thence, and also to use them as flankers, he contrived to make communications, like bridges, from the wall of his intrenchments to the roofs of the houses; and having barricaded the doors and windows of the ground floors, he posts guards in them; but these thefe precautions were ufelefs. The Huffars, tired with watching him, fell back towards their army; and M. de Clairac retired to Tirs-chen-reit, whither he was going.

Thefe examples, which prove what great refources a well-informed genius will derive from courage, fhew alfo to what a pitch the defence of an intrenched houfe may be carried, under the conduct of a determined refolution. The only means whereby the enemy may eafily force it, is to batter it down with cannon. If they once take this method, I fee no poffibility of their holding out long, unlefs, after it is quite down, they can contrive to range themfelves about the intrenchments.

Peafants houfes are generally fo ill built, and every cannon-ball is likely to make fo great a breach, that the defenders muft expect, in the end, to be buried under the ruins. The only methods then to efcape perdition, is either to capitulate, or to fally out brifkly on the enemy when he leaft expects it. The firft method cannot be thought of,

only

only upon condition of obtaining the honours of war; which are, to march out by beat of drum to return to the army, whither you are to defire to be efcorted and conducted by the fhorteft road *. If the enemy will grant no capitulation, as the condition of foldiers, prifoners of war, is always more grievous and unhappy than death itfelf, fo death would be preferable to it, if one had not ftill the refource left, which experience proves to be almoft certain, of faving yourfelf by a bold fally. The neceffity one is then under of conquering, transforms the brave man into a defperado, and opens him a paffage either to his army, or to a neighbouring poft. It was by a *manœuvre* of this nature, that Count Saxe (afterwards Marfhal-General of the French armies) efcaped from Crachnitz, a village in Poland, where 800 of the enemy's horfe

* To ftipulate to be conducted by the fhorteft road is a very material article; a General, who forgot this circumftance in his capitulation at Oftend laft war, was marched half round Flanders with his garrifon, inftead of being conducted by the fhorteft road.

horse set upon him, and 18 of his followers, with intention to take him. This Prince resisted them a long while in the chambers of an inn at this place, when seeing himself unable to hold out any longer, he sallied out unexpectedly in the night, sword in hand, cut his way through the midst of the enemy's guards, and retired to Sandomir, where he had a Saxon garrison.

When the resolution is taken to abandon a post, which can be no longer maintained, you should continue a very smart fire till the instant the sally is made; and, in the mean while, remove with the least noise possible, the barricade from the door by which you are to issue; when that is done, assemble speedily all your party on the ground floor, march out in the closest order possible, and with fixed bayonets drive rapidly towards that part you have perceived to be least guarded. "One should
" never wait for day-light," says the Chevalier Follard *, " to make these
" sallies, which can only be succefsful
" in

* Comment. on Polyb. Tom. V.

"In a dark night, by the opportuni-
"ty it affords of concealing from the
"enemy the road by which the retreat
"is made." For this reason you must
clear your way with your swords, and
not suffer a shot to be fired, lest the
enemy should direct their strength to
the place where they hear the noise.

M. the Baron de Travers gives us
also a good lesson on this subject. "To
"avoid being met by the enemy," says
this author*, "one should always take
"a road quite different from that
"which the enemy might suppose we
"did take, and which should appear
"to be what we ought to take: a small
"party can hide themselves any where;
"and as it is not common to seek those
"places near the enemy, those there-
"fore are the safest; there they should
"pass the day, and take another road
"under cover of the night."

But if the post be considerable, such
as a village or borough, whereof the
defence is committed to an officer, he
may kill a great many of the enemy

before

* Etudes Militairis.

before he is obliged to make his retreat. When his smaller posts have held out as long as possible, he will make them fall back to his principal one, still fighting from street to street, and from trench to trench. But in order that the solders may execute these *manœuvres* easily, he must, as I have before said, have exercised them therein before-hand. In a defence of this nature, the commandant should also observe with great attention all the motions of the enemy, so as to distinguish a feint from a real attack.

Though the enemy should succeed so far as to force all the intrenchments, and to get footing in the village itself, it must not therefore be taken for granted, that he has gained the victory. An officer, retired to his principal post, may begin a-new to give him such a reception, as I have particularized in the method of defending single houses, so as to disgust him entirely with his enterprise, and oblige him to retire again.

I have

I have said, and shall repeat it again, the defence of a post, of a village, or even of a city, is so easy, that I cannot comprehend why they do not hold out longer than they commonly do. There is nothing necessary for it, but resolution, vigilance, to know how to make the most advantage of the ground, and to persuade the soldiers that nothing but downright cowardice can let the enemy penetrate into the place. The example of Cremona in 1702, will be an everlasting proof of what determined courage can do, and will teach posterity that, though the enemy should be possessed of half the ramparts, and of a part of the town itself, they are not yet entire masters of the place.

Prince Eugene having formed a design to surprize this town, which was our head-quarters, defended by a garrison of French and Irish; some thousand Austrians were introduced there by a priest. These troops immediately made themselves masters of two gates, and of a great part of the town: the garrison buried in sleep, started up in surprize, and

and were obliged to fight in their shirts; but the French officers directed their *manœuvres* with so much prudence, that they repulsed the Imperialists, from place to place, from street to street, and obliged Prince Eugene to abandon the part of the town, and the ramparts that he was in possession of.

What then hinders us, now-a-days, from defending villages where we are posted, in like manner, and from disputing the ground inch by inch; especially when a church, or *chateau*, is secured as a sure retreat, fit to make a good defence, and to obtain an honourable capitulation? This is easy, and yet we see few or no examples; because we do not apply ourselves sufficiently to learn the causes of the disasters, which our predecessors fell into, for want of knowing better.

One may judge, from what I have said on the defence of posts, that there is nothing more easy than to maintain them; those who attack, having nothing supernatural in them, but are the
same

same sort of men with those who are attacked.

A determined commander, who is jealous of his reputation, and who has learned, by study, to make use of his talents, dares, like Leonidas, with 300 men, defend the streight of Thermopile, against a whole army; and as a modern philosopher says, had rather perish gloriously, than be guilty of a cowardly action. In fact, an able commander is never astonished at the numbers of the enemy; in a house, a village, or even in a town, he may oppose devices to them, which will always supply his defect of forces.

I have seen, during the last war in Piedmont and Italy, intrenchments and posts that have withstood the first and the strongest attacks of the assailants; and which have been given up or abandoned in some following attacks, tho' not near so hot as the former: from whence comes this? It is because the officer that is placed in the post, dare not abandon it on the first attack; but he defends himself and repulses the enemy;

enemy; because if he suffered himself to be forced, he and his men would be all put to the sword. But should the enemy return, a commandant imagines that he has nothing to reproach himself with, because he defended it some time; and then he either retires, or surrenders.

But since he was able to repulse the enemy immediately when they came fresh, and in good order, to attack him; with how much more facility may he repeat the same, when they return harrassed and fatigued, and in a condition much less to be dreaded than the first time? Does not the cause of this come from, not exciting sufficiently the emulation of military persons? An officer who is not countenanced, and who is never assured of the least reward, neglects himself, and thinks less of acquiring glory, which, though obtained by a brave action, is for the most part unknown, than of enjoying quietly a common reputation.

"We commend greatly," says Monsieur Follard*, "and I believe we "can-

* Commentaries on Polybius, Tom. 5.

" cannot too much either commend or
" reward those who make a noble de-
" fence in the posts that are committed
" to their charge. The reason of this
" is, that the recompences for these
" kind of actions being much greater
" than those given for others, excite
" and animate officers to defend a post
" vigorously to the last extremity.—
" But if the requital should be pro-
" portioned to the action, then he that
" has done nothing worthy of a brave
" man, but has surrendered in a cow-
" ardly manner, ought to be stripped
" of his arms, and put to death with-
" out mercy. This was a law with
" the Romans *; but the General, on
" his part, should be attentive to leave
" the officer no room to complain, and
" that he should be furnished with eve-
" ry thing requisite for his defence.
" It is not necessary, says the same au-
" thor in another place, that an officer
" who is fixed in a post should be over-
" ready

* And in France also; by an ordonnance of the 20th of July, 1714, it is forbiden, under pain of death, to quit or desert a post,

" ready to come to action; but that he
" shall always resist when he is pressed,
" and that he should die, rather than
" abandon his intrenchment."

Ancient and modern history furnish but few examples of posts being well defended; and it seems as if the military authors had agreed not to speak of actions of this nature. Nevertheless, it is not to be doubted, but that in the different wars which France has waged, chiefly under the reign of Henry the Fourth, and of his Successor, in times when the armies were not near so numerous as they are now-a-days, there were officers whose actions deserved to be remembered in history. However, we don't find any body has recorded them; although the lessons that might have been taken from them, would have been as instructive and as agreeable to read, as those that have been left us of the best fortified places of a state If I am surprised to see in 1604 a hundred thousand men perish before Ostend, and their General *,
with

* Archduke Albert.

with the remains of his army, not able to mafter the place, till after a fiege of three years; I am no lefs aftonifhed to fee Charles XII. King of Sweden, in 1713, with feven or eight officers, and fome fervants, defend a wooden houfe near Bendar againft twenty thoufand Turks or Tartars. We find in many hiftorians the defence of this houfe, becaufe it was made by a crowned head; but great actions, let the authors be who they will, ought not to be buried in oblivion. They not only pique the emulation of officers, who always find matter of inftruction in them; but they are honourable to thofe who perform them, to the corps they belong to, and even to their nation. I am forry I could not collect a greater number of authentick ones, for it would have been a great pleafure to me, to have embellifhed this little work with them.

X CHAP.

CHAP. X.

Of the attack of posts.

THOUGH to take a post from an enemy may be always a difficult task, if those who are to defend it know their business; however, the way to succeed is, either by a rapid and sudden attack, or by stratagem.

One ought never to form the project of an attack upon simple speculation; because our imagination often makes us think things feasible, which, when we come to the proof, we find impossible to execute. When an action of this kind is proposed to be undertaken, one should form a just idea of it; examine separately every part of it, and the various means that are to be made use of, and compare them together, to see if they correspond with each other, and answer to the general end: lastly, one should take such just measures, that, if I may be allowed to say so, you may be insured of success before you begin.

As it is not customary in an army to chuse a particular officer of foot to attack an intrenched post, unless he offers his service; an officer ought not to embark in such an enterprise, without examining into the means of succeeding, and being able to shew a plan of his project to the General, in order, if he approves of it, to gain his consent to put it in execution. In case the General likes the plan, the officer is to desire leave to take a nearer view of the post, with one or two men, in order to take more exact measures for the execution. I say he ought to ask leave to go and view the post, to the end that if he should be discovered and taken prisoner, he may be owned, and reclaimed.

How to reconnoitre a post.

An officer who goes to take a near view of the post he intends to attack, should go out at the beginning of a dark night, and give those, that go with him as assistants, instructions how to act in every circumstance; such as, to examine well every place through which they pass; to approach the post, by searching with long poles, lest there should be any traps or ditches covered over, into which they might fall; and to stick large branches of trees, with the leaves on, at those they discover, so as to guide them when they return to the attack; to take particular notice of the position of the centries, their distances from each other, and their number; to advance to the edge of the ditch, to try the depth of the water with their poles, or with a lead and line; to see whether the post is *fraised* * or palisaded, built of earth or fascines, or masonry: in the last case they are to guess, as near as they can, at its height, to be able to proportion the length of their scaling-ladders. Lastly, to know how many men the garrison consists of, and in what they are negligent; if they are

* A *fraise* is a pallisade laid horizontally, or nearly so, being half buried in the earth of the parapet; the other end sharpened, and pointing out towards the enemy Pallissade, properly so called, is commonly set perpendicular.

are likely to receive any succours, or if they have any cannon, &c.

It is upon the knowledge of all these circumstances, which one can examine into himself, or may learn from the report of deserters or peasants, that an officer may form the project of an attack.

If a person receives his instructions only from the reports of others, he must be cautious how he gives credit lightly to those, whom perhaps either a desire of betraying him, or the hopes of recompence, might have induced to throw themselves in his way; on the contrary, he should question them separately, write down what they say, compare their depositions, and judge afterwards what part of their intelligence may be true, or what false. Having taken these instructions, the officer should return to his General, to inform him of his discoveries, and receive his last orders for the attack; for the soldiers that are to second him, and those who are to march to support him.

Of the choice of soldiers.

The choice of men who are to march to the attack of a post is a thing so essential, that the success of the enterprize depends upon it. Therefore none should be chosen but willing and bold soldiers, and who are not rash or heedless, and none who have colds; for a man, who without attention to the orders of his officer, will suffer himself to be led by the heat of his zeal, or who by coughing or spitting, discovers the march of his party to the enemy's centries, may make the best concerted project fail. As to those that are to support him, he is to take them according to their turn for guard, or for detachment, as the General shall think proper.

Of Dispositions.

The dispositions for an attack ought to be relative to the discoveries that have been made; so that one should not be obliged to return in the middle of the execution.

The men being chosen, they are to be inspected, to see that they want nothing that may contribute to their success; I say, that may contribute to their success: because, if the post is fortified with an intrenchment of earth, or of *fascines*, the two first ranks should be provided, besides their arms, with shovels and pickaxes; if it is *fraised* and pallisadoed, they should have good hatchets; and if it is faced with stone or brickwork, they should carry scaling ladders. All the soldiers also ought to be in their waistcoats, so as to be more at liberty; and they should have paper cockades, that they may know one another in the dark: after this inspection, the following disposition is to be made.

If the intention is to make one or two true attacks, and as many false ones, the chosen men are to be formed into as many platoons as there are to be true attacks; and the others appointed to support them, are to make the false ones, in order to divide the enemy and their fire. Then a man, capable of commanding, should head each platoon; and observe ferve that thefe officers, as far as poffible, may be the fame who were at the examining the fituation with him; fo that each of them may be able to guide his divifion.

Thefe officers are to be ordered to march together, till they come to the place appointed for their feparation; whence they are to go each to their feveral ftations near the poft, where they are to lie down on their bellies, and wait for the fignal of attack to leap into the ditch, and fcale the poft.

Of Guides.

If an officer is to be conducted to a poft, by guides or fpies, he fhould firft queftion them carefully, fo as to draw from them as much ufeful information as he can, particularly touching the nature of the road, by which they intend to conduct him. The reafon of which is, becaufe one fometimes meets with filly fellows, who, animated with the love of lucre or otherwife, believe that they can conduct a body of men

eafily,

eafily, at the fame time that they are totally incapable, and truft folely to their good will. But if he finds them fufficiently qualified for the purpofe, he fhould ufe all poffible means to be affured of their fidelity, by making them dread the total deftruction of their houfes, and pillage of all their goods, if they lead his troops into any traps. He may alfo demand their wives or children, as pledges for their good behaviour; and at the time of marching, they fhould be placed between the corporals of the firft rank, fecured with a fmall chain or cord. This laft precaution is the more neceffary, as traitors have been often found, who under pretence of helping to furprize a poft, have conducted a body of men into a cutthroat place in the middle of the night, and flipped off themfelves in the midft of the fray. On one hand, a recompence proportioned to the fervice fhould be offered to thofe people, in cafe their conduct fhould be good; on the other hand, they fhould be threatened with the moft fevere punifhment if it fhould be bad.

Of the March.

Night being the beſt time to ſteal upon a poſt, care muſt be taken to ſet out time enough to get near the place an hour or two before day, provided that it is not moon light at the ſame time; for if it be, the attack ſhould be deferred, if poſſible, till the moon gets under a cloud, and then ſeize the moment of obſcurity to begin the work: the ſoldiers ſhould march, two by two, as lightly and ſilently as poſſible, eſpecially when they are to paſs between two of the enemy's centries, they are to be forbid to talk, cough, ſpit, or ſmoak.

When the detachment is arrived at the place where the platoons are to ſeparate, the officers of theſe diviſions are to repair, with them, to the places I before-mentioned, for them to lie down, to wait the ſignal; obſerving, that the places where they lie in ambuſh, ſhould be oppoſite the ſalient angles of the intrenchments, ſeeing that theſe are the

ſpaces

spaces least defended by the enemy's musketry.

If whilst the parties are on their march, or lying in ambush, they should happen to meet the enemy's patrole, they are not to be alarmed on that account, or make the least motion, because it might spoil the whole enterprize: they should only remain hidden, and totally silent; so that the patrole may pass by without seeing them, and that they may afterwards pursue their own design.

The attack of a common redoubt.

If the post to be attacked is a redoubt, with a dry ditch and a parapet of earth; the two first ranks of each division, as I said before, are to be provided with shovels and pick-axes, and are to sling their muskets. Things being thus prepared, as soon as the chiefs see, or hear the signal, all the divisions are to rise, and march with speed, to leap into the ditch at the same instant; I say at the same instant, because it should be a maxim in the attack of a post, for all to fall on at once. When the firſt rank have leaped into the ditch, the ſecond ſhould ſtop a moment, leſt they ſhould leap on the backs of the firſt, and throw themſelves on their bayonets. Theſe two firſt ranks being got into the ditch, they are immediately to undermine the angles of the *ſcarpe* or ſlope, and the parapet of the redoubt, to facilitate the climbing up of the reſt of the party. The officers of each diviſion are to take care, that the ſoldiers armed with their muſkets, who have alſo got into the ditch, may not obſtruct the workmen, but that they protect them, by preſenting to the right and left; and that they are always ready to repulſe the enemy that may have been poſted in the ditch. If the parapet be *fraiſed*, they are to cut away with their hatchets, as many of theſe pointed poſts as will leave a ſufficient paſſage; and when the breach is made, the workmen are to lay down their tools, handle their arms, and mount all at once with fixed bayonets, and fall upon the enemy, crying out, kill, kill.

When

When a body of men march against a redoubt, or any other post, with intent to surprize it, the commanding officer ought always to make his attack on that side, which may have communication with some other more considerable posts, in order to cut off this communication; for people who see themselves warmly attacked, and have no hopes either of retreat or succour, will very soon ask for quarter.

The attack of intrenchments with a revêtement, i. e. faced with masonry.

Though the attack of posts, whose *scarpe* or slope,* and parapet are faced with brick or stone-work, can only be made by *escalade*; nevertheless it succeeds, if it is briskly surrounded, and well supported.

An officer who intends to attack a post of this kind, must take care that the

* The *scarpe* or slope, reaches from the bottom of the ditch, up to the ordinary level of the ground; and the *parapet* is the super-addition lying above the *berme*, which marks the surface or level of the ground.

the ladders he intends to use, shall be rather too long than too short; and let only the strongest soldiers carry them. These soldiers are to carry them with their left arms thrust through the second step; they must keep them upright, and close to their sides; and hold them so short, as not to come near the ground, to avoid dislocating their shoulders, when they leap into the ditch. The first ranks of each division, being thus provided with ladders, are to set out at the first signal, with the rest of the party, and are to march boldly with their pieces slung, and their swords in their right hands, and leap into the ditch: when they are down, they are to fix their ladders against the wall, rather towards the salient angles than against the curtins; because there the enemy are by much the weakest; and care must be taken to fix the ladders but a foot from each other, and not to give them too much or too little slope; as in one case they may be easily overturned, and in the other they will be too weak to bear the men.

When

When the ladders are fixed, thofe that carried them are to mount directly, and to be followed immediately by all the reft, to fall on the enemy fword in hand. If he that gets up firft fhould be knocked down, the next man muft take care not to let himfelf be beaten down by the falling body; to avoid which, he muft endeavour to make him pafs on one fide, between the two ladders, and then climb up as quick as he can, not to give the enemy time to reload.

As the foldiers who mount firft are likely to be knocked down by the enemy's fire, and as their falling may happen to make the attack fail, I think it would be right to give them a light cuirafs or breaft-plate; becaufe, as foon as thefe have got in, the reft will eafily follow. Some people may think this an ufelefs precaution; but is it better then to have all your people expofed to be knocked on the head in the ditch, than to carry the poft with more certainty and lefs danger?

The

The success of an attack by *escalade* is infallible, if they mount with speed on the four sides; if they take care to shower in grenades in abundance; and if they are supported by some companies of grenadiers, and by some pickets, who will draw off and divide the enemy's fire and attention.

Of the passage of a ditch full of water.

If the ditch of the post to be attacked is full of water, and only takes a man up to his belly, that need not hinder their jumping into it, and carrying on the attack, as described in the foregoing section; but if it is too deep to be passed by wading, the soldiers of each platoon must carry fascines or faggots, of slender branches, made as thick as possible, and tied very tight, to fill up the ditch, or render it so far fordable, that the assailants may get to the parapet, either to undermine it, or to scale it.

Some authors recommend, for this purpose, casks filled with earth; and

M. de

M. de Follard, facks filled with dung *, or litter, of five feet diameter; but I have found, by many trials, that the cafks are very difficult to roll, especially if the ground is uneven, and you have a confiderable way to move them; that it is difficult to fill up the ditch with them, because their folidity makes the water rife higher and higher: that facks of earth, or of dung, cannot be rolled, on account of their weight; that they burft in the carriage, fpill their contents, make the ford very muddy, raife it but little, and leave it ftill difficult to be paffed. Therefore fafcines are preferable to all thefe, becaufe the foldiers can carry them before them, where they ferve to cover them from mufket-fhot; and being light, they do not retard their march. All thefe fafcines, which may be handed from one man to another, and thrown into the water, will foon fill up the ditch, fo as to make a paffable ford.

M. Follard gives us another method of paffing a wet ditch; which is, to make

* Tom. III. pag. 408.

make frames of seven or eight feet broad, by ten or twelve long. "These frames," says he *, "consist of three bars of "wood, with cross-bars, in the manner "of a hurdle, and well mortoised; "planks should be nailed on the top, "and a grapple fastened to one end, "to cling to the fascines of the enemy's "intrenchment."

But M. Follard has not told us how these bridges are to be carried to the ditch, or how an officer is to get them made.

Ways to counter-act the other contrivances.

If the approaches of a post are defended by *chevaux de frise*, the first and second ranks of each platoon should cut away the spikes with hatchets, or they may hale them forward, and throw them aside, with an iron grapple fastened at the end of a rope; if by an *abbatis*, they should throw fascines or great faggots on the points, and over the branches, by which means the soldiers will

* Commentaries on Polybius, tom. V.

will be able to pafs eafily over it: alfo, if thefe *abbatis* are double or triple, they may be fet on fire, by throwing well dried faggots, lighted at one end, into the middle of them. If this laft propofition is to be executed, as foon as the lighted faggots are thrown on the *abbatis*, the foldiers fhould retire to a certain diftance, fo that the enemy may not have an opportunity to level their fhot at them by the light of the fire; and they alfo fhould be fo pofted, as to be able to fire at thofe of the enemy who fhall endeavour to extinguifh the flames.

But, laftly, if the avenues to a poft are defended with caltrops, they muft be fwept away, by dragging one or two trees, with all their leaves on, over the ground where they have been laid.

The attack of a chateau, or of a houfe.

The approaches of a *chateau*, or a houfe, are to be attacked in the fame manner as thofe of a detached or fingle poft: after thefe are taken, trial is to be made to gain the upper part, by escalade, and to destroy the besieged by knocking them down with tiles: but if the enemy have uncovered the post, in order to prevent this; then hand grenades should be thrown in at the door and windows as fast as possible; also a quantity of dry fascines, with lighted faggots dipped in rosin, should be thrown in, to endeavour to stifle or burn out the enemy. If it should be windy weather, advantage may be taken thereof, to blow the flames towards the house; and attempts should be made to stop the loop holes the enemy may have made in the walls with sand bags, in order to set about *sapping*, or undermining the angles of the building; but if the assailants have any cannon, the work may be shortened, by pointing them at the angles of the post. Instead of cannon a great beam may be suspended with ropes from a triangle, made with three strong poles, in imitation of the *battering ram* of the Romans. This beam forcibly pushed against the walls, will soon make a large breach;

but

but care muſt be taken only to mount it in a dark night, ſo that the enemy may not obſtruct the work, dy killing the men that are employed about it. If it is a glorious part to come off with honour from an attack of this kind, it is not leſs ſo to execute it with the loſs of only a few men. One cannot be too ſaving of the blood of ſoldiers; and a knowing officer ought never to neglect the means that can contribute to preſerve his men. The compariſon of two examples that I will relate, will ſhew the importance of what I advance. During the two ſieges of Barcelona made by M. de Vendome in 1697, and M. de Barwick in 1713, the firſt of thoſe generals attacked ſword in hand the convent of capuchins ſituated without the town, with ſeveral detachments of foot, and made himſelf maſter of it in three hours time, with the loſs of ſeventeen hundred men. Marſhal Barwick attacked the ſame convent in 1713: the enemy were equally intrenched, and reckoned to make the French pay as dear for the victory as they had done before; but
this

this general opening a kind of trench before the convent, the enemy, who did not expect to be attacked in form, surrendered at discretion, after twenty-four hours resistance. I submit it to the judgment of my military readers, to decide which of these examples is best to follow.

The attack of a village.

The preparations for the attack of a village, or any other extensive post, are the same with those mentioned in the former part of this chapter for those posts that stand alone. But as these kinds of attacks are always more difficult than others, on account of the various devices that may be opposed to every attempt; an officer should not begin his movements,* till he knows the strength of the intrenchments, the situation of the little posts, the obstacles

* That is, he should not move until he has got the best information the nature of the thing admits of; an officer must be cautious in construing this sentence, not to draw hence reasons for his inactivity, under pretence of his not having sufficient intelligence.

stacles he may meet with in every street, and how the inhabitants stand affected towards the garrison.

If an officer takes this information from the people of the country, he should affect a great indifference in his enquiries, that they may not suspect his design, and communicate it to the enemy, who by that means may take precautions to overturn his project. He should also endeavour to be well assured of their truth, as I said before, by comparing the reports of peasants, deserters, and what he knows, or has seen himself, all together, in order to find out the most probable. When he knows the enemy's situation, he is then to make his dispositions for the attacks, and must point out the duty to the officers of each platoon, as well to those of the false attacks, as to those of the true.

The true attacks are to be made at the places that are most difficult of access, because here the enemy, confiding in the strength of the situation, are the
least

least on their guard *. They may also be made upon the houses situated at the entrance of the streets, because, when you are in possession, it is an easy matter to break through the walls of one house into another; and being possessed of the houses, it will be easy to drive the enemy out of the streets even with stones, and oblige them to take to their last intrenchment.

If the war is in an enemy's country, which you do not chuse to spare, it would be an easy matter to set fire to the four corners of the village, and oblige the besieged to surrender themselves presently; but besides the inhumanity of using means that tend to lay waste a whole country, it is likewise very dangerous to throw all the inhabitants of the open country into despair, because then flying into the woods, they form into bodies, and spread about every where,

to

* People always guard against probabilities; therefore, as Cardinal de Retz says, probabilities seldom come to pass:—at the late conquests of Cape Breton and Quebec, the successful landings were made at places, deemed by the enemy (the French) to be inaccessible.

to knock the straggling soldiers on the head, murder the suttlers, hinder the peasants to carry any provision to the camp, and ravage the whole country. M. Follard * speaking on this subject, says, "we saw during the war in 1688, "fifteen hundred Barbetts of the valley "of St. Martin, keep forty batallions "of our troops in awe, through the "whole extent of the valley of Pra-"gelas, where the Bisan runs in the bot-"tom between two very high moun-"tains, of most difficult access, and "where each party guarded their own "side." .... These mountaineers came down sometimes when they saw our convoys in motion, and attacked them. At the same time they had scarcely ten or twelve men where we had entire corps.

It appears from the example I have now mentioned, that it is very imprudent to master considerable posts, such as villages, by setting them on fire; and that it is much better to take them by smart attacks.

A a

An

* Commentaries on Polybius, tom. IV

An officer who commands an expedition of this kind, should not attach himself obstinately to one single attack; for the false ones may become true, and he ought to know the success of each, so as not to throw away men to open a passage on one side, whilst perhaps it may be already open on the other.

When the assailants have penetrated the village, the commanders of each division should be attentive to leave small detachments at every church, and at every strong and tenable place fit for the bulk of their party, in case they should be repulsed.

They should be very watchful that the soldiers do not scatter about to pillage the inhabitants houses. Detachments have been often driven back out of a town or village, by neglecting this precaution.

Three days after the surprize of Cremona, in 1702, some German soldiers were still found in the cellars, where they had got drunk, and were greatly astonished, when they were told that they must quit those lovely retreats.

An

An officer, who would avoid so dangerous a disorder, should make it death for a soldier to stir from his party, and should post serjeants in the rear of each division, to prevent any man from staying behind.

If they find any cavalry drawn up in the markets, or open places of the village, the besiegers are to stop, and stand firm at the heads of the streets opening into the place or market: some of them are to get into the houses next the corners, to fire on the enemy through the windows; and if they find it disorders them, they should immediately charge them with fixed bayonets, to oblige them to surrender.

Lastly, if the internal parts of the village be defended by cannon, the troops should march with speed to possess themselves of them, to spike them up, or to turn them on the enemy, or against their chief post in the village.

One may judge, from all that I have said on the surprizing and seizing of posts, that those actions, tho' difficult, are not impossible, when the means intended

tended to be employed in the execution are judiciously combined, and well considered. These ways are easy to be imagined; nevertheless few examples of such actions appear, because people do not apply themselves enough to this part of war; wherein to succeed, it requires good sense, great courage, a head for stratagem, a daring spirit, a ready execution, and a cautious foresight.

We find in antiquity, an example of an attack, which by the recital of its circumstances, may be of great service to regimental officers: I take it from the seventh book of Polybius. "The "blockade of Sardis by Antiochus the "Great," says the translator * of that author, "held for two years, when La-"goras of Crete, a man skilled in the "art of war, put an end to it in this "manner" He had remarked that the strongest places were often taken with the most ease, owing to the negligence of their defenders, who confiding in the strength of the natural or artificial fortifications of their town, take very little

* Don Vincent Thuillier.

pains

pains to guard it. He knew alſo that places were ſometimes taken by an attack on their ſtrongeſt ſide, and where the defenders did not expect that the enemy would undertake any thing.

And though he ſaw plainly that Sardis was looked upon as an impregnable fortreſs, not to be attempted by aſſault, and that nothing but famine could oblige them to open their gates; nevertheleſs, upon the foregoing conſiderations, he hoped to ſucceed. Difficulties only increaſed his application to think of all poſſible means of entering the place; and obſerving that the part of the wall which joined the citadel to the town was not guarded, he formed a deſign to ſurprize it in this place; the proof of this ſide's not being guarded, was thus: this wall was built on a very high and ſteep rock, at the bottom of which was a kind of deep pit, wherein the town's people were uſed to throw the dead carcaſſes of horſes and beaſts of burthen; there vultures and all ſorts of carniverous birds aſſembled every day, and after eating their fill, never
failed

failed to rest themselves either on the rock or on the wall.

This was enough for our Cretan engineer, to see that this place was for the most part of the time neglected, and without any guard. Upon this supposition, at night he went to the place, and examined how he could approach it, and where he might apply his ladders. Having found a fit place against one of the rocks, he imparted his design and his discovery to the king; who, delighted therewith, exhorted Lagoras to go on with his enterprize, and gave him two other officers that he required, and who seemed to have all the abilities and valour that his project demanded. These three having consulted together, only waited for a night, at the end of which there should be no moon light. When it came, they chose fifteen of the strongest and stoutest soldiers in the army to carry the ladders to scale the walls, and to run the same hazards with themselves. They also chose thirty others to lie in ambush in the ditch, to assist those who should scale the

the wall, to break open a gate to which they were to march. The king was to order two thoufand men to follow thefe, and to favour the enterprize by marching the reft of his army to the oppofite fide of the town. Every thing being ready for the execution, as foon as the moon was down, Lagoras and his people approached foftly with their ladders, and having fcaled the rock, they came to the gate which was near it, and broke it open, having forced all whom they met in their way; the gate being open, the two thoufand men entered, when fome cutting down all that oppofed them, while others fet fire to the houfes, the town was pillaged and ruined in an inftant.

I intreat young officers, who fhall read this example, to confider well on this attack. The diligence and attention of Lagoras, who went himfelf to view the proper place to fix his ladders, his difcernment in his choice of the officers and foldiers that were to aid him, and the exact concord between the feveral means that he ufed, are fo many leffons
for

for those who shall be tempted to undertake such an attack.

That of M. de Roche-Fermoy, captain in the regiment of Bourbonnois, performed before Charleroy, is no less instructive.

During the siege of that place, at the end of July, 1746, M. de Lautrec, then lieutenant-general of the trenches, perceiving that the taking the redoubt of Mareinelle, which defends the lower town, was absolutely necessary to streighten the garrison; he ordered M. de Roche-Fermoy, a brave and determined officer, to take a close view of this important post. This officer accepted the commission, set out with a single man at the beginning of the night, passed between the enemy's centries, and advanced even upon the glacis of the redoubt. Having sounded the water, which seemed very deep, he found in one place five or six feet, and in another only four. He also saw the post was *fraised* and palisaded, and defended by several pieces of cannon, and fifty Austrians commanded by three officers.

officers. All these obstacles were not sufficient to dishearten him; he tied his cockade to some reeds opposite the place where the water was but four feet in the ditch, and in his way back left his coat at some distance from that, to be a guide to him on his return to attack the post. When he came back to the trenches, he reported his discoveries to M de Lautrec, who gave him forty chosen men, and ordered him to be supported by Monsieur de la Merliere, captain of grenadiers, who having marched to the opposite side of the redoubt, and drawing the fire of the garrison that way, favoured this enterprize. As soon as M. de Roche-Fermoy got near the glacis, he made all his men lie down close, to wait the signal, which was made an hour before day. Then this officer leaped into the ditch, ordering his men to put their carteridge boxes on their heads, to preserve them from wet; made them cut away with their hatchets as many of the *fraises* as were necessary to open a passage, climbed up the redoubt, and fell on the enemy with fixed bayonets, who being surprized at so sudden a visit, sought their safety by flight; but M. de Roche-Fermoy having ordered the draw-bridge, that communicated with the town, to be hoisted up, they were obliged to surrender at discretion. The smartness of this attack, the orders of M. de Lautrec to make it succeed, and the activity with which the prince of Conty directed the works of the siege, making the enemy fear their being taken by assault, the Generals Beauford and Halkert, who commanded, gave up the place to the prince, and were made prisoners. The next day M. de Roche-Fermoy was presented to his highness, with all the Germans he had taken; the prince greatly commended his valour, and sent so favourable an account of his conduct to court, that the king immediately settled a pension upon him, to be paid out of the royal treasury.

I confine myself to the examples that I have given, on the subject of *attacking posts*, having found no more among several that I have looked over, that were sufficiently authentic, or well enough explained to make use of; intelligent officers will supply this defect by their own reflection. Those who really love their profession, will both gain knowledge, and make new discoveries by application only. Certainly I should be flattered, if what I have given should be approved; but much more so, if the rules that I have proposed can contribute to the success of officers, so as to distinguish them, that they may arrive at those high ranks which are the recompences of military virtues.

Of seizing posts by stratagem.

That part of military science which comprehends the surprizing of posts, is little capable of being treated in a methodical manner. The particular intelligence of each officer, and several opportunities that happen by chance, are what commonly occasion the execution of these kind of actions.

War being an art depending much on *fineſſe* and ſtratagem, there are a number of precautions that eſcape the foreſight of men in action, which a ſkilful enemy may obſerve, and which furniſh him with opportunities to make good ſtrokes History contains many examples of the like actions, which are only rare now-a days, becauſe we do not ſufficiently ſtudy this part, wherein is required an elevated genius, and a combination of means relative to the true poſition of the enemy, which one ought always to view one's ſelf.

But alas! how many people loſe their underſtanding when they are obliged to take a cloſe view?

I diſtinguiſh two ſorts of ſurprizes in war, the general, and the particular. The firſt are ſuch as are undertaken againſt a camp, an army, or a fortified town. As it is neceſſary, in order to ſucceed in theſe, to take the precautions that have been already laid down by able heads, and to be aſſiſted by a much great number of men than what private officers commonly command; I

ſhall

shall not speak of them. I shall confine myself to those that may be performed by a small body of men, or for the success of which very extensive measures will not be required.

In all times stratagems were used in war, "says the author of the military "dictionary, a work useful for all of-"ficers. Frontinus under Trajan, and "Polyenus under Antoninus, have "wrote on the stratagems of great cap-"tains, and even of illustrious women. ".... Each general has his own, adds "this author; there are some which "owe their rise to time and place only, "and which ought never to be neg-"lected. Many people pretend that "every thing is fair in war, and that "you may procure the success of what "you undertake, by any means what-"soever. But authors, who have wrote "on the *jus gentium*, or the rights of "nations, do not agree on this head."

I will remark on this subject, that M. le Chevalier Follard thinks all stratagems equally good; tho' in the number of those that he relates, there are,
according

according to him, some wherein sincerity and greatness of soul shine with great lustre; and others where the most infamous treachery, and the most cruel practices, are only looked upon as the *finesses* of a knowing enemy.

"This part of war, says this great
"author,* has not been sufficiently
"explained; it were to be wished,
"however, that these kind of works
"were often read, and considered by
"persons of the profession. This read-
"ing appears to me to be the more use-
"ful; for, besides that it is amusing,
"it is also instructive; and that when
"one knows the stratagems and devices,
"he knows how either to render them
"useless in the enemy, or to employ
"them himself when he has occasion."
To which I will add, that those opportunities often occur in the space of a campaign; and that they offer every day without being perceived, for want of attention; nor are they known till after they are passed by. However, nothing contributes more to the reputation of an

* Notes on Polybius, tom. IV. p. 30.

an officer than thefe kind of actions; but for that, fays Vegetius, you muft lend a hand to fortune, and know how to profit by the advantages fhe offers you.

Among the furprizes of pofts that may be executed by a fmall body, and which an officer may undertake; there are fome, to which we are invited by their apparent facility, or by our watching the enemy with great attention.

I will not repeat on this fubject what I faid before, of the precautions to be taken in marching towards a poft; it is enough to know that as thefe projects are like machines, where the difplacing a fingle wheel makes all the others ufelefs; you muft examine with the greateft nicety, every way and means intended to be ufed, to fee that they correfpond well with each other, in order to fucceed in the meafures you have taken. How courageous foever the officers in our army may be, it muft not be imagined that they are all capable to embark in enterprizes of this kind; for befides that they fhould be impene-
trably

trably fecret, and of great difcernment in the choice of their foldiers; they fhould alfo have a perfect knowledge of the country, and be able to fpeak the language; but few officers are poffeffed of all thefe qualifications

As to the manner of furprizing a poft, I have faid that it was impoffible to eftablifh certain rules on this fubject, becaufe among a thoufand opportunities that chance offers to us, there may not happen to be two alike. A very quick march ftolen upon a diftant poft, where the guard is negligent; a thick fog which prevents them from being feen; a river that has a ford unknown to the enemy; an aqueduct, or fubterraneous paffage, or a hollow way that is not guarded; a little river frozen over; a road ftopped up; good intelligence; the time of a fair; market-day; and difguifes of all kinds. Such are the different ftratagems which may be ufed occafionally, and even promife a happy iffue, though the fame have been often made ufe of. I will only obferve, that there are ftratagems where

it

it will be impossible to succeed, unless you join therewith a steady force. A considerable post, such as a town or village, for example, where a party is to be introduced by having an intelligence within, cannot be carried, unless one is well seconded. The only means to conduct the surprize of these posts well, is to make it a rule to divide your force, to make yourself master of the *chateau*, the church, church-yard, and all the public places. It is a mistake to say that troops, divided in this manner, could only act weakly, and would run a risk of being knocked on the head one after another. I would chuse always to make as many detachments as the enemy had posts; because in the fright, caused by these surprizes, it is easy to make yourself master of those posts, before their defenders have time to dispute them, or even to look about them. The enemy being obliged to divide their force also, and not knowing to what place to give the preference, it is almost a moral certainty, that being stunned with the noise on every side, they will even drop their arms out of their hands: let us add alſo, that the horrors of a dark night, and deſpair that never fails to poſſeſs a body of men that are ſurprized, repreſenting objects much greater than the reality, they will imagine they have a whole army to encounter with.

The ill ſucceſs of the ſurprize of Cremona in 1702, where the Germans had divided their forces, proves nothing againſt my opinion. If without ſtopping to make priſoners, they had march'd a detachment directly to the caſtle, which ought always to be an object of principal attention in thoſe kind of actions, it would have been impoſſible for the brave officers, who repulſed the Imperialiſts, to have made ſo glorious a defence. M. de Schover, who ſurprized Benevare in Spain, in 1708, took quite a different way, and accordingly gained his point. This General having learned that the Spaniards neglected the guard of an old caſtle, which was at the entrance of that town, marched thither in the
night,

night, and made himself master of it, and afterwards sent sundry detachments to attack the Spaniards in the town; but these surprized by so sudden a visit, endeavouring to save themselves by flight, ran to the castle, as the last resource of a garrison; where, as fast as they entered, they were made prisoners. This method, to go strait to the castle of a town that one intends to surprize, is then the best to follow; because the enemy do not expect that the first attack will be made upon their strongest part; and being in equal concern both for the town and citadel, it is to be presumed that they will have divided their forces, so as to be able to defend all parts alike.

If any thing has been the cause that these events are not frequently seen now-a-days; it is, that they are almost always forgotten, and that the authors of them scarce ever obtain any recompence; nevertheless, what does not a man merit, who determines on an action of this kind, bravely adventuring himself to be sacrificed in the attempt?

M.

M. Menard has given us a relation of a surprize,* in the history of the town of Nismes, which deserves to be told, for the good lessons it contains. Nicholas Calviere, called Captain St. Cosme, having resolved to make himself master of this town, agreed with a miller, whose mill was within the walls, near one of the gates, called *de la Bouquerie*, that he should file, during several nights, an iron grate, that shut up the entrance of the aqueduct, thro' which the spring-water passed into the town; that he should cover the parts cut by the file with wax, so that they might not be perceived in the day time; and that he should conceal a hundred armed men in his mill, while a more considerable body of horse and foot, that were to arrive from Vivarais, advanced to support the enterprize.

The orders for the rendezvous of the troops being given, and the day for the execution fixed to the 16th of November, 1569, St. Cosme sallied out of the mill

* History of the town of Nismes, tom 5, in 1569, and note 2d.

mill at three o'clock in the morning, advanced to the guard-houfe of *la Bouquerie*, killed all the foldiers, and opened the gate for two hundred horfemen, each of which carried a foot foldier behind him. Thefe men being got into Nifmes, immediately formed into feveral detachments, one whereof went to block up the caftle, whilft the others, going to all the fquares and open parts of the town, and founding their trumpets, became mafters of it in an inftant.

I thought this furprize of Nifmes the more neceffary to be related here, as its circumftances are very inftructive. Captain St. Cofme, who knew how to profit by the governor's neglecting to guard the entrance of an aqueduct; the choice he made of horfe to expedite the coming of his foot from different quarters; the exactnefs of his orders to thefe troops, which were about fifteen leagues from Nifmes, to be at the place of rendezvous at the hour appointed; his precaution in fending to inveft the caftle, to avoid having the garrifon to fight with in the ftreets; his attention

to

to distribute his forces to all parts of the town, and to make them sound trumpets, to make the inhabitants think their numbers were very great: all these circumstances are so many instructions for officers who may be tempted to undertake a like enterprize.

M. Carlet de la Roziere, engineer at the isle of Bourbon, has collected some examples of surprizes, which have succeeded by a particular conduct. Brachio, a captain of Jean, queen of Naples, being desirous to become master of a tower in the territory of Ambresa, disguised one of his soldiers like a woman, giving him a basket in his hand, together with a sickle. This man, so transformed, ran as fast as he could towards the tower, feigning to fly for fear of some of the enemy's parties; the guard let him in, and even let him go up a ladder to the top of the tower to shew the centry where the enemy were. But as soon as he got up, he clove the centry's skull with the sickle, and seized his arms, with which he drove off those that were below from their post.

It

It is in this manner, that sometimes where it is impossible to conquer by force, the mind should be attentive to profit by the least fault of the enemy. Epaminondas knowing that his wife was beloved by Phebiades, governor of Cadmia, ordered her to sup with the governor in the citadel, which he defended, and to invite a great many other ladies. The wife obeyed, and the guests came to the rendezvous. But towards the end of the repast, those ladies going out of the citadel to a nocturnal sacrifice, which was only to last a few minutes, the guards were ordered to let them pass. As soon as they were out, Epaminondas ordered them to lend their cloaths to some chosen soldiers, who being introduced to the citadel by one of these ladies who had the watch word, surprized the governor, and became masters of the fortress.

Necessity is in war, as every where else, the mother of invention, when one has firmness enough not to be discouraged. The Amphictions besieged Cirrha; the greatest part of the inhabitants

bitants were supplied with water from a plentiful spring, by means of an aqueduct. Eurilochus, * one of the Generals, having discovered this aqueduct, he ordered a great quantity of hellebore to be mixed with the water; scarcely had the defenders of Cirrha drank of it, when they were tormented with such dreadful gripings, that they were incapable of defending themselves; so that the Amphictions became masters of the place without the loss of blood.

" The disproportion of forces," says M. le Chevalier Follard, somewhere on the subject of surprizes, " is not always
" in the number, but often in the ca-
" pacity of the one, opposed to the ig-
" norance or neglect of the other."
Mary, queen of England, not being able to reduce the duke of Suffolk, the head of the party that disputed
the

---

* Frontinus and Polyænus disagree on the subject of the author of this device; Frontinus attributes it to Colixthenes, also General of the Amphictions, and Polyænus to Eurilochus.

the crown with her in 1553,* sent one hundred bold soldiers to the fortress where he was intrenched. These men reported themselves to him as deserters, that were coming to join his party; but as soon as they got in, they turned their arms against him, and seizing his person, gave him up to Mary, who beheaded him.

The civil wars of France, about the end of the sixteenth century, produced also some examples of posts being carried by stratagems: captain Martin, and du Rolet, governor of Pont-de-l'Arche, having formed a design on Louviers in Normandy, in 1591, surprized that town by means of a corporal, a priest, and a tradesman. The priest took upon him to keep watch in the belfry, † and to let the troops advance

---

* This story rests entirely on the credit of some French writer, whence most probably M le Cointe has copied it; our histories, I believe, say nothing like it,

† It is always customary, especially in time of war, to keep a watchman, both day and night, on church steeples, to give notice of the appearance of troops, or of fire.

vance as near the town as they pleased, without ringing the alarm, and the two others promised to deliver up the gate. These measures being taken, Du Rolet sent forward seven resolute soldiers with black scarfs, which was always worn by those of the league; these stopped under the gate of the town, where the corporal and the tradesman talked with them, as with people of the union. Du Rolet being informed by the tradesman that it was time to fall on, came out of his ambush, ran to the gate, took possession, cut the guard in pieces, entered the town, and immediately became master of it, with the assistance of new troops brought to sustain him by the Baron de Biron.

There are none but those who have the strongest desire for glory, and whose valour is never diminished by danger, who know how to reduce an enemy by stratagem, and to seize an opportunity that fortune offers to them.

Gustavus Vasa, seeing the sea frozen over, crossed it with his army, and went in the middle of the night to burn the naval

naval army of the Danes, that were stopped at a little distance from Stockholm, whither they were going to increase the power of tyrants, and the despair of the people.*

I will add but one word more on the subject of surprizes, on which one might write volumes; it is, that after one has formed a design, and examined all its branches very well, one should never stop in the middle of the execution, for having discovered an obstacle that was not foreseen. Dionysius having intelligence in the town of Naxos, appeared before the place at night with a considerable body. The garrison being informed of the treachery, took up arms and got on the ramparts; Dionysius, though astonished, was not discouraged; on the contrary, he threatened to put all to the sword; and sent a boat into the harbour with a certain number of boat-

* Amazing! that such an expression could fall from the pen of a French soldier, when he and all his companions have been fighting for a century past, only to swell the vanity and importance of the tyrants that enslave them.

boatswains, (or those who keep the galley rowers to their work,) with their whistles; and as each gave his signal different, the Naxiens imagined that there were as many galleys in the port as they heard whistles, and surrendered at discretion. If Dionysius had endeavoured to retreat on seeing his design discovered, he would have been greatly exposed: for besides, that the Naxiens could have come out, and cut off his rear guard, he would have been also the object of their railleries. After the battle of Cannæ, Hannibal advanced as far as the gates of Rome, with a design to lay siege to it. But he was hindered by a great noise he heard in the night, like a number of people laughing very heartily: the Romans being astonished the day after at his retreat, built a temple immediately, which they dedicated (Deo ridiculo) to the god of laughter.

I will say no more on the seizure of posts by stratagems, because one may see by what I have said, that these actions in general are not so difficult as they are commonly thought to be. Timorous

morous people, and thofe who in the fmalleft affairs are ftopped by the moft trifling difficulty, may probably look upon them as impoffible, and even think there is fomething fupernatural in thofe they fee fucceed; but it is not for fuch people I write. To men of application, of bravery, and bright genius; and in a word, to the officers of our own times, who are now in fervice; to them I fubmit my ideas, and them I have chofen to judge of the methods I have herein propofed.

The end of M. le Cointe's treatife.

Monsieur de la Croix, who published a small treatise in 1752, intitled, *Traité de la petite guerre*, or a treatise on petty war, gives the following account of his design.

"THE different actions in which I have been engaged, during fifty years that I have had the honour of serving the king; the important expeditions of which I shared in the execution with my late father, *(a major general)* have given me experience in the *petite guerre*: I have endeavoured to learn every thing that could contribute to raise the service of *free companies* * to the greatest degree of perfection: I have laboured to remove inconveniencies, to reform abuses, and to establish an order and discipline suitable

and

---

* A *free company* is one of those corps commonly called *irregular*, is seldom or never under the same orders with the regular corps of the army, but for the most part acts like a detached army, either by itself, or in conjunction with some of its own kind; therefore their operations are properly considered under the title of the petty war.

and proper for that *corps*; I have taken pains to gain the confidence both of officer and soldier; to know their qualities and their virtues, in order to employ them usefully. Succefs in divers encounters has taught me how to attack, and how to avail myfelf of ftratagems of war on all occafions that occur, and in all conjunctures and countries that I happen to be in. I thought I could not make a better ufe of my little knowledge than to publifh it, in order that thofe who enter into the military profeffion may thereby improve themfelves to the advantage of the fervice."

But as a great part of M. de la Croix's treatife feems to be a pious panegyric on the actions of his father, the tranflator propofes only to extract a few paffages, and fuch as may confonantly be added to the foregoing treatife.

De la Croix's precautions for a march.

Stratagems are to be employed in all enterprizes againft the enemy; but it is is also essential to dive into those that they may use against you, so that you may be able to elude and counteract their effects.

When any important and distant expeditions are in agitation, and when it is required to send great detachments so far as sixty leagues or more, you should begin by taking the best instructions you can, relative to the country through which you are to pass, what routes you are to take, the situation of the enemy, of the strength and situation of the different posts possessed by them; and on your departure from your garrison, you are to see that every man comes out with every thing necessary for the expedition; (but in order to deceive the enemies spies,) you may form many small detachments of twenty-five, thirty, or forty men, to the number of two or three hundred, which may march out by different gates, and on different days.

These divisions are each to be headed by an officer, who being informed of the march of the other detachments, must

muſt regulate the days of his march, ſo that they may all meet nearly at the ſame time at the place appointed for their re-union, by a certain ſignal agreed on. Here a general inſpection is to be made of the whole body, to ſee that none have deſerted; after which they are to march, by filing off in great ſilence, and keeping the by-roads; the villages ſhould be avoided, and they ſhould only march at night, and not exceed four leagues at moſt; at daybreak they ſhould throw themſelves in ambuſh, into a wood, to wait till night, and refreſh themſelves with the proviſions that they took care to furniſh themſelves with before their march. But theſe precautions, however wiſe and prudent they may be, are not near ſufficient for the ſecurity of a body that are to penetrate into an enemy's country; the commanding officer ought not only to be careful to make the march ſilent and ſecret, but he ſhould foreſee how he is to return: he ſhould never move one ſtep till he has projected the means of his retreat: the execution

ecution of his design ought to engage him to keep up a most discreet and modest behaviour towards the country people wherever he is to pass, and to make the march as little burthensome to them as possible; he must gain their esteem by his affability, good order, and the sobriety of his men. This must be followed by some presents prudently bestowed, which will not only prevent his being harrassed on his march, but will dispose the country people to discover some important particulars to him, and shew him the motions of the enemy, their numbers, position and strength. These instructions being confirmed by spies, and the people that he has sent before to examine the country, will enable him to act with some confidence, and make him almost sure of the success of all his enterprizes.

To prevent detachments from being discovered by the barking of dogs, when they are passing by farm-houses that lie separated from villages, the officer should order some men disguised to go before, furnished with poisoned pills,

pills, or *nux vomica*, for those animals. These people serve as well as spies to make discoveries of consequence, for the security of a march.

The use of infantry, and the utility of horse in a retreat.

All considerable expeditions are made by the foot. The horse intended for this service are not to set out till some time after the foot, they are to approach near the place where the blow is to be struck, to lie in ambush till the moment of its execution; and then, for the most part, have no more to do than to sustain the foot in their retreat, which is to be made by long marches; and in order to make this retreat more secure and easy, they should endeavour to procure, in the places they pass through, a sufficient number of good waggons and stout teams of horses, which helping the foot to forward their march more briskly, may also serve as a barricade against the enemy's horse that might pursue them into the plains.

M.

M. de la Croix alluding to his father's and his own past experience, shews how the most difficult things used to be executed.

The most difficult projects were formed and were executed; difficulties never discouraged us, all obstacles were surmounted, and the enterprize had a happy issue. The reason is plain, the troops were experienced and accustomed to war; the officers had good understanding, and were men of honour; the commanders in chief were assured of the merit of those they employed; their method also was not to be rash, knowing that immoderate heat and too much precipitation, instead of advancing the success, would make it miscarry; and that by trusting too much to chance, the best enterprizes may fail. They knew how to temporise wisely, to allow their project time to ripen. They had trusty people in different places, whom they rewarded with great punctuality, who furnished them with exact accounts. Their least step and all their motions were

were guided by prudence; their operations had always a succefsful end. What better maxims can be laid down in the military art?

Care and precautions to be taken in towns, villages, and places of refreshment.

A body or detachment are conducted as their chief thinks proper, and put an implicit confidence in him, when he has given them proofs of his vigilance and attention to their safety. Therefore when he enters a village or town to refresh, he should immediately post double centries in the steeples, or highest buildings, which are proper to make discoveries from, to observe the environs, to prevent surprizes and unforeseen attacks; then he is to distribute the provisions, and give out the necessary orders to his people; he is not to confine himself to this alone, he must artfully pick up useful and necessary intelligence; he must talk with the burgomaster and other principal people of the place, endeavouring by obliging

means

means to gain their confidence, to draw from them some interesting confessions; he is to demand of them trusty persons to send before him, and to promise to pay well for any services that they may do for him: lastly, he must spare neither money or pains; the money is most efficacious, it must be liberally disposed of on proper occasions, and without regret; and the returns will be ample in the advantages that will result from it.

Other precautions and measures for night marches; attention to the fire arms; and the essential custom for retreats.

I have said before that night is the best time for a march; and it cannot be secret at any other time; but great care must be taken in the dark. A body should file off slowly, regularly, and in silence; the commander should order halts from time to time for indispensable necessities, and order the officers to watch while the men are marching in a file, lest they mistake one road for another, and to remain in the rear till such
time

time as they are again rejoined in a body or column. No man muſt be ſuffered to ſmoak, even in the ambuſcades, on account of the inconvencies of the ſmoak and the ſmell of the tobacco: if they are to paſs through plowed ground they ſhould drag large faggots of briars after them to efface the marks of their feet, left the peaſants ſhould obſerve them: when they arrive in a wood at day-break, as the leaves are commonly covered with dew or rain, thoſe at the head ſhould carry a kind of blinds of oiled cloath to cover them, to break way for the reſt to follow, ſo that they ſhall not be wet.

To keep the fire arms in good order, and to prevent them from being wet in rainy weather, the method is to have a ſmall caſe to draw over the butt to cover the lock of the piece; this is ſoon pulled on and off, and will keep the lock and priming in order to fire, which is not always the caſe; for how many regiments when it rains, march without attention, and out of order, the ſoldiers carrying their butts behind expoſed to the rain, and

may

may be attacked by a much lefs body ? examples of which have been feen. This caution is not all; the ferjeants fhould vifit the men's arms every day; and as ammunition is in one refpect more precious than provifions, it fhould be managed with the greateft care. A foldier ought to have at leaft a hundred rounds to fire; and he fhould never difcharge one without effect, as *free companies* have no train waggons to attend them.

Laftly, as we ought to forefee every circumftance, we fhould be provided againft every accident; we fhould carry grenades, combuftible ftuff, caltrops, nails to fpike up cannon, petards, hatchets, fhovels, match, and jointed harrows: the ufe of thefe things is foon known, they ferve to burn forage, and hinder and delay the purfuit of cavalry in a retreat.

An ufeful maxim for rencounters, nocturnal and unforefeen attacks.

All the devices and precautions ufed by the commander of a body, are intended

tended to perfect the project he has formed; this same reason should make him attentive to attempt nothing on the enemy that may thwart or delay the execution of that project. Sometimes an opportunity offers to fight, and even crush a detachment passing by his ambuscade; but he must be careful not to take this advantage, lest he thereby obstruct his project; and if by accident he should meet a body at night, he should proceed thus, in order to gain his point. His advance guard should be preceded by two or three men, who should go on very silently, and stop now and then to listen; if they hear any thing, they should come without noise to give an account thereof; but if they should unexpectedly fall in with the enemy, they should call out loudly *who comes there?* At this noise the body fix their bayonets, keep close, and throw themselves to one side, on the right or left of the road, to wait the result. The commander attentive to the motions of the enemy, who, as it often happens, might have only come towards him by chance, lets them pass by; but if they come on imprudently, tho' they may be superior in numbers, he is in a condition to receive them with his body, whom he makes kneel with fixed bayonets. However, these are accidents which it is always prudent to avoid; for though one may come off victorious from such an encounter, yet the body is weakened thereby, and these kind of successes often deprive you of the power of executing the only project proposed.

A stratagem commonly made use of by M. de la Croix.

The orders being given to set out a detachment of three or four hundred men, they marched by divisions, as beforementioned, of thirty, forty, or sixty men, and by different roads, but to meet on a fixed day; then they kept close in woods, or where they could lie hid; and a body of sixty or eighty men were detached to go and refresh themselves in the neighbourhood of the place where they wanted to allure the enemy:

the

the commander or governor of the neareſt town, on hearing of this body's approach, from the inhabitants of the place, never failed to ſend out a large detachment, in proportion to the numbers he heard were in the country; many have been very roughly treated who did not expect to be attacked by more than the reports gave out, till they were drawn into the ambuſcade by the alluring party.

The advantage of night attacks, and the precautions to be taken in quarters.

Night attacks are almoſt always ſuccefsful, and the reaſon is pretty evident. The aſſailants are informed of the poſition and the ſtrength of the enemy; the latter are ignorant both of the numbers and of the manœuvres that are to be employed againſt them; the one knows where to ſtrike, and is ſure of his blow; the other hardly knows what part he is to defend: in theſe circumſtances whole battalions have been beat and routed by moderate detachments. There are
some,

some, who confiding in their numbers, and the valour of their men, and satisfied to be told that there is no considerable body of the enemy near them, abandon themselves to their ease, and cannot be persuaded that two or three hundred men could come to insult them; in this false opinion, as soon as they arrive at a town or village, the commander, after having ordered the quarters, appointed the posts, and placed the guards, seeks a good lodging for himself, and gives himself up to his ease; the other officers follow his example, and take care to want for nothing comfortable; and all indulge effeminately in the middle of danger: but they often pay very dear for such imprudent conduct; the enemy, who are on the watch, are informed of their arrival; spies bring them news of the true state of things every where, and they soon become acquainted with the position of the advanced guards, and of the commander's quarters.

These kind of enterprizes have always been looked upon as very bold
and

and even rafh, to dare to attack a body of fix or feven hundred men, with a detachment of two or three hundred; yet it is not to be doubted but a true partifan, who is well acquainted with the country, and with the march of a fuperior body, may eafily form his attack in the dead of the night, and better in bad weather than in good, as he has his arms always dry by the method I mentioned before, let the weather be as it may; on fuch an occafion he arrives at a village, with his party at the diftance of a league or more from the enemy, where, during his halt, he informs himfelf, by the chief magiftrate, of every particular, who will not difobey him; he is alfo to afk for fome men of the place to ferve to help him to reconnoiter the enemy; fuch people are always to be found, who for a proper recompence, or from an inclination to be contrary to different troops, are eafily determined to this fervice; they are to be inftructed what to do, and what to obferve; to know where the guards are pofted; where the commander is lodged;

if

if there are no ways of surprizing them by going behind through some gardens; he should ask them if they have got any relations in the places to name, in case the guards should stop them, and so take off all suspicion. After these measures, they are to be ordered to return to an appointed place, when they are ready to make their report. Those expeditions seldom fail; and to succeed, the body is to be divided into three or four detachments, with a view to fall on all at once, and not to give the enemy time to look about them; but should any one say, what confusion at night? How can these detachments join again? The answer is, that truly these attacks are very hazardous to both one side and the other; but the assailants are never embarrassed for the following reason, which is easy to be conceived, that before the attack they take care to send eight or ten soldiers, each carrying a truss or two of straw on a stake, to set fire to at the moment of the attack; this fire serves as a direction to those who attack to retire to the light after they have

have taken some prisoners. All these kind of attacks are made in less than half an hour, and the enemy cannot know the meaning of the fires; and this device hinders them from observing those who attack them, so as to be able to pursue them in their retreat.

A commanding officer cannot be too circumspect in and about his quarters, especially while ever he is an enemy's country, where the natural aversion of the inhabitants will be joined to the activity of the enemy to harrass and overpower him.

The secret of marching small divided bodies of a detachment, and to make them rejoin quickly at the appointed time and place, is of infinite advantage, and puts them in a condition to form their attack with more certainty of success, as the enemy don't expect to have to do with a large body; they are not concerned when they hear of forty or fifty men only in the field: if they are even told of another body of the same number, that have been seen, they are persuaded that it is the same they heard of before; and they are seldom undeceived, till the time that the union of the whole is made, and ready to begin an attack, which they never apprehended.

End of the extracts from M. de la Croix.

Some

Plate 1st

Plate 2.d

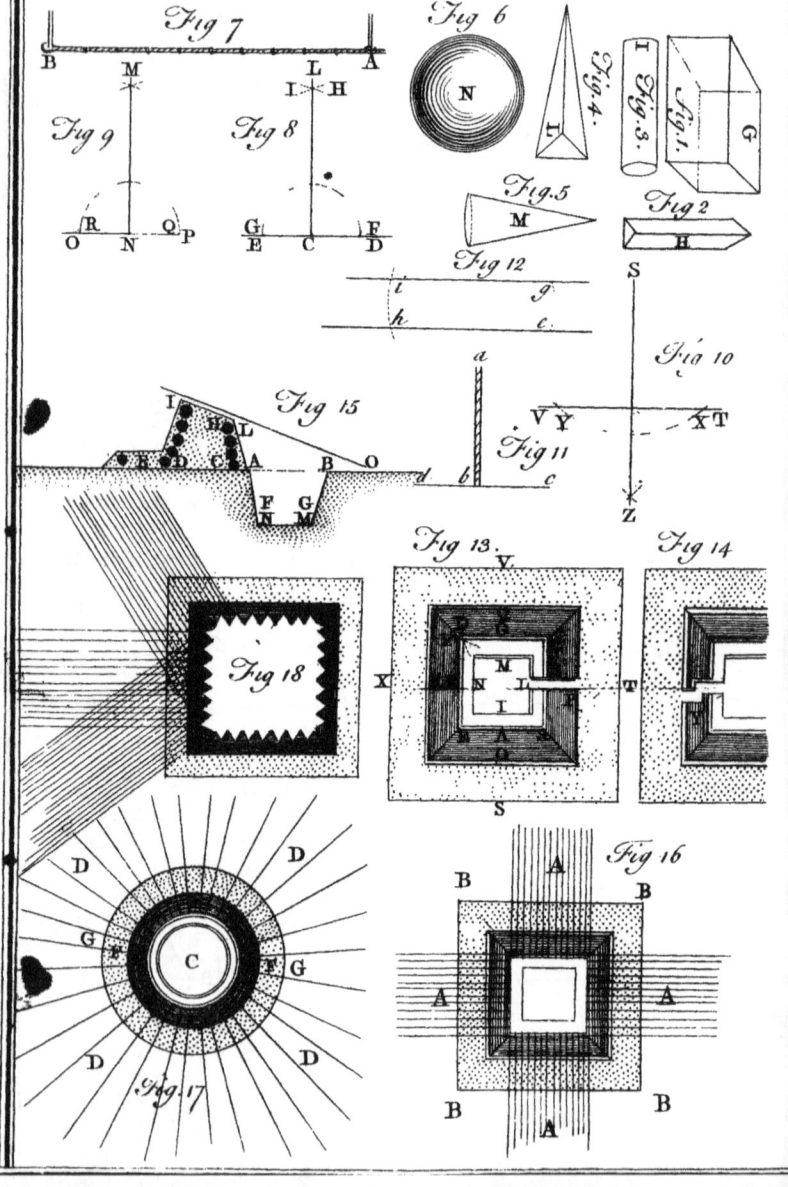

Plate 3ᵈ

The View of a Redoubt

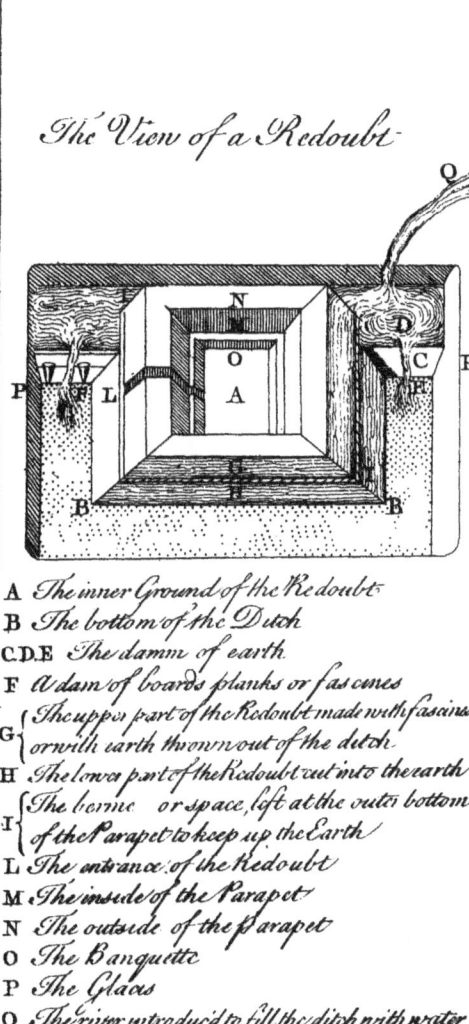

A *The inner Ground of the Redoubt*
B *The bottom of the Ditch*
C.D.E *The damm of earth.*
F *A dam of boards planks or fascines*
G { *The upper part of the Redoubt made with fascines or with earth thrown out of the ditch.*
H *The lower part of the Redoubt cut into the earth*
I { *The berme or space, left at the outer bottom of the Parapet to keep up the Earth*
L *The entrance of the Redoubt*
M *The inside of the Parapet*
N *The outside of the Parapet*
O *The Banquette*
P *The Glacis*
Q *The river introduc'd to fill the ditch with water*

[225]

Some hints and observations borrowed from Marshal Saxe.*

Of war among mountains.

THERE are few things to be said on this subject. But those who make war in mountains should be extreamly cautious; they should never venture into the valleys without first possessing the heights, then all ambuscades are at an end, and they may pass on

---

* The translator is sensible that the foregoing extracts, as well as these hints borrowed from M. Saxe, relating more to the duty of a *general* than of a *private* officer, rise above the plan laid down by Monf. le Cointe, and may seem improperly added to such a work; but as in the British service every subaltern should expect to rise to the highest military rank, he thought it not amiss to open the prospect at the end of M. le Cointe, and to give them a small view of the path they are to pursue in their studies for the higher offices, whereby they may see that the duty of a *general* officer is only the exercise of the *machine* at large, whereof that of the *subaltern* is the *model* in miniature.

on in safety; without this precaution there are great rifques of being knocked on the head, or of being obliged to return, after lofing a great many men.

But if you find the paffages, as well as the heights pre-occupied, you fhould make a feint to force them, to amufe and attract the attention of the enemy, and at the fame time feek a pafs through fome other way. Though difficult the mountains may appear, one may always by ftrict fearch find paffages. The people who inhabit them do not know them themfelves, becaufe they have not been obliged to feek them; therefore one fhould not believe their reports, who, for the moft part, are acquainted with the things relating to their country by tradition only: I have often found out their ignorance, and the falfity of their relations.

In fuch a cafe one fhould fee and fearch one's felf, or employ people who will not ftop at difficulties; thefe things are always found when diligently looked for; and the enemy, ignorant thereof, runs away, not knowing what courfe to take,

take, becaufe he only provides for common things, that is, againſt the moſt practicable roads.

Of a country inclofed by hedges and ditches.

As the enemy is as much embarraſſed in this kind of country as you can be, there is not much to be apprehended; thefe are little matters that decide nothing; and where the moſt obſtinate will fucceed. There is only one thing to be minded; it is to keep all behind clear, fo as to be able to detach, and to retire, in cafe of neceffity. Here it is very neceffary to know how to place cannon, which is of great fervice: as the enemy dare not ſtir from the poſts they poffefs, one may cannonade them with eafe; if they abandon them, their retreat is not always fuccefsful, and one has fometimes the good luck to cut them off.

But as I faid before, that thofe things in the whole are not very decifive, they fhould be guided by the fituation of the places; fo that no certain rules can be prescribed. Nevertheless one should always observe this, to push forward, and on your flanks in marches, with detachments of one hundred men, supported by double, and the double by a triple, so as to be covered and in safety.

A detachment of six hundred men will stop an army; because on the highways that are inclosed by hedges and ditches, such as are found in Italy, and in all well watered countries, one shews a great front to the enemy, who will suppose your numbers to be much greater than they are. The least hut makes a fortification where one may support a very obstinate engagement, which will give one time to look about, and to make a disposition; for we should guard against surprizes in these kind of countries.

A partisan of spirit and address, with three or four hundred men, will cause a frightful disorder, and will attack an army very well on its march: If he cuts off the baggage at the beginning of the night, he will carry off a great part of it, without running much risque; because if he retires between two ditches, and secures his rear, by blocking up or otherwise embarrassing the road, he will stop you; if he is pushed, he leads the waggons in a line, and the first house he finds, he stops you short, during which time the baggage he has taken advances in the country. If he acts thus against your horse, he will put them into great disorder; it is for this reason that you should have advanced, and flanking parties in and about all the avenues of your march, and they should not be weak; for there is no doubt but the enemy will be lying in wait for you on such an occasion; and you must fight to the last minute to avoid dishonourable surprizes. If you have to do with an enemy, whose general has common sense, he will soon find people in his army, who have bold and penetrating minds, so as to see things just as they are.

Of the paſſage of rivers.

It is not ſo eaſy as one imagines to hinder the enemy to paſs a river; but he can come to attack you much eaſier, than he can defend himſelf when he is retiring before you. In one of thoſe caſes he ſhews you his front, and ſupports himſelf by a good diſpoſition, and by a hot cannonade; but in his retreat he ſhews you his rear, which is very difficult to defend, and much more ſo as he is hurried; and as this diſpoſition is never ſo well made as that of an attack; and as every body in a retreat contracts a kind of timidity, which makes them feel already half vanquiſhed.

In regard to the paſſage of a river by main force, I believe it is not poſſible to prevent it, eſpecially when it is ſupported by a briſk cannonade, which gives time to the head to intrench themſelves, and to make a work to cover the bridge. There is nothing to be done

done in the day-time, but at night one might attack this work; and if it should happen to be at the time that the enemy's army begins to pass over, confusion spreads every where, and those who have got over already will be lost; but one should attack with strength; and if you let the night slip, you'll find all the enemy passed over the next day. Then it is no small action, but a general one, that is to be risked, and which reasons of state do not always permit.

F I N I S.

www.ingramcontent.com/pod-product-compliance
Lightning Source LLC
Chambersburg PA
CBHW031140160426
43193CB00008B/196